Uncertainty and Operations Research

For further volumes:
http://www.springer.com/series/11709

Meilin Wen

Uncertain Data Envelopment Analysis

Springer

Meilin Wen
Science and Technology on Reliability
   and Environmental Engineering Laboratory
School of Reliability and Systems Engineering
Beihang University
Beijing, Beijing, China

ISSN 2195-996X ISSN 2195-9978 (electronic)
ISBN 978-3-662-43801-5 ISBN 978-3-662-43802-2 (eBook)
DOI 10.1007/978-3-662-43802-2
Springer Heidelberg New York Dordrecht London

Library of Congress Control Number: 2014944128

© Springer-Verlag Berlin Heidelberg 2015

This work is subject to copyright. All rights are reserved by the Publisher, whether the whole or part of the material is concerned, specifically the rights of translation, reprinting, reuse of illustrations, recitation, broadcasting, reproduction on microfilms or in any other physical way, and transmission or information storage and retrieval, electronic adaptation, computer software, or by similar or dissimilar methodology now known or hereafter developed. Exempted from this legal reservation are brief excerpts in connection with reviews or scholarly analysis or material supplied specifically for the purpose of being entered and executed on a computer system, for exclusive use by the purchaser of the work. Duplication of this publication or parts thereof is permitted only under the provisions of the Copyright Law of the Publisher's location, in its current version, and permission for use must always be obtained from Springer. Permissions for use may be obtained through RightsLink at the Copyright Clearance Center. Violations are liable to prosecution under the respective Copyright Law.

The use of general descriptive names, registered names, trademarks, service marks, etc. in this publication does not imply, even in the absence of a specific statement, that such names are exempt from the relevant protective laws and regulations and therefore free for general use.

While the advice and information in this book are believed to be true and accurate at the date of publication, neither the authors nor the editors nor the publisher can accept any legal responsibility for any errors or omissions that may be made. The publisher makes no warranty, express or implied, with respect to the material contained herein.

Printed on acid-free paper

Springer is part of Springer Science+Business Media (www.springer.com)

Preface

In the past 30 years, data envelopment analysis (DEA) has grown into a powerful quantitative, analytical tool for measuring and evaluating relative efficiency of decision-making units (DMUs). DEA has been successfully applied to many different types of entities engaged in a wide variety of activities in many contexts worldwide. Although DEA offers more advantages than many other statistical approaches, some limitations have to be considered. One important problem is its sensitivity to data. Therefore, a key to the success of the DEA approach is the accurate measure of all factors, including that of inputs and outputs. However, in many situations, such as in a manufacturing system, in a production process, or in a service system, inputs and outputs are so volatile and complex that they are difficult to measure in an accurate way. Thus, various uncertain DEA methods have been proposed to handle the data variation in DEA. This book is intended to present the milestones in the progression of uncertain DEA.

Uncertain Theories

Based on different uncertain theories, uncertain DEA methods deal with the problems involving uncertain inputs and uncertain outputs. Chapter 1 will present the basic introduction to uncertain theories which are crucial for the following chapters, including probability theory, credibility theory, uncertainty theory, and chance theory.

Introduction to DEA

Chapter 2 is intended to represent a milestone in the progression of DEA, including a comprehensive review and discussion of basic DEA models and extensions to the basic DEA methods.

Stochastic DEA

Incorporation of random variations into DEA analysis has received considerable attention in recent years. Chapter 3 will present some stochastic DEA model extensions of the usual deterministic DEA formulations. This kind of models makes it possible to replace "efficient" and "not efficient" with "probability efficient" and "probability not efficient," respectively.

Fuzzy DEA

Fuzziness is a basic type of subjective uncertainty, which has been applied in DEA recently. Chapter 4 will provide some fuzzy DEA methods based on credibility measure, including fuzzy DEA models, fuzzy sensitivity analysis, fuzzy fully ranking methods, and fuzzy congestion.

Uncertain DEA

Uncertainty theory is a branch of axiomatic mathematics for modeling human uncertainty. Chapter 5 will apply uncertainty theory to DEA so as to produce a new method of dealing with the empirical data. Besides uncertain DEA models, uncertain sensitivity analysis, uncertain fully ranking methods, and uncertain congestion will also be taken into account.

Hybrid DEA

In order to evaluate the DMUs in which uncertainty and randomness appear simultaneously, Chap. 6 will introduce a hybrid DEA method based on chance measure. Hybrid DEA models, hybrid fully ranking methods, and hybrid congestion will be included in Chap. 6.

Purpose

The purpose of this book is to provide some methods in dealing with uncertainty in DEA. The book presents four uncertain DEA methods based on different measures including probability measure, credibility measure, uncertainty measure, and chance measure. The book is suitable for researchers, engineers, and students in the fields of management science, economics, operations research, industrial engineering, information science, and so on.

Acknowledgment

This work was supported by the National Natural Science Foundation of China (No.71201005).

Beijing, China
October 1, 2013

Meilin Wen

Contents

1 **Uncertain Theories** 1
   1.1 Probability Theory 1
       1.1.1 Probability Measure 1
       1.1.2 Random Variable 2
       1.1.3 Operational Law 4
       1.1.4 Expected Value 6
   1.2 Credibility Theory 11
       1.2.1 Credibility Measure 11
       1.2.2 Fuzzy Variable 17
       1.2.3 Expected Value 22
   1.3 Uncertainty Theory 23
       1.3.1 Uncertain Measure 24
       1.3.2 Uncertain Variable 24
       1.3.3 Operational Law 27
       1.3.4 Expected Value 31
   1.4 Chance Theory 35
       1.4.1 Uncertain Random Variable 35
       1.4.2 Uncertain Random Measure 36
       1.4.3 Operational Law 38
       1.4.4 Expected Value 40
   References 44

2 **Introduction to DEA** 45
   2.1 Symbols and Notations 46
   2.2 CCR Model 46
   2.3 BCC Model 50
   2.4 Additive Model 53
   2.5 SBM Model 56
   2.6 Russell Measure Model 57
   References 58

3 Stochastic DEA ... 61
- 3.1 Symbols and Notations ... 62
- 3.2 Stochastic DEA Models ... 62
  - 3.2.1 Marginal Chance-Constrained Models ... 66
  - 3.2.2 Satisfying DEA Models ... 72
- 3.3 Stochastic DEA Ranking Criteria ... 76
  - 3.3.1 Expected Ranking Criterion ... 76
  - 3.3.2 Optimistic Ranking Criterion ... 77
  - 3.3.3 Maximal Chance Ranking Criterion ... 78
  - 3.3.4 Hurwicz Ranking Criterion ... 78
- References ... 80

4 Fuzzy DEA ... 83
- 4.1 Symbols and Notations ... 83
- 4.2 Fuzzy DEA Models ... 84
- 4.3 Sensitivity and Stability ... 88
- 4.4 Fuzzy DEA Ranking Criteria ... 94
  - 4.4.1 Expected Ranking Criterion ... 94
  - 4.4.2 Optimistic Ranking Criterion ... 95
  - 4.4.3 Maximal Chance Ranking Criterion ... 99
  - 4.4.4 Hurwicz Ranking Criterion ... 100
  - 4.4.5 Numerical Examples ... 104
- 4.5 Fuzzy Congestion ... 106
  - 4.5.1 Congestion in DEA ... 106
  - 4.5.2 Fuzzy Congestion ... 108
  - 4.5.3 A Numerical Example ... 110
- 4.6 Hybrid Intelligent Algorithm ... 111
  - 4.6.1 Fuzzy Simulations ... 111
  - 4.6.2 Genetic Algorithm ... 112
  - 4.6.3 Hybrid Intelligent Algorithm ... 114
- References ... 114

5 Uncertain DEA ... 117
- 5.1 Symbols and Notations ... 117
- 5.2 Uncertain DEA Models ... 118
- 5.3 Sensitivity and Stability ... 122
- 5.4 Uncertain DEA Ranking Criteria ... 128
  - 5.4.1 Expected Ranking Criterion ... 128
  - 5.4.2 Optimistic Ranking Criterion ... 130
  - 5.4.3 Maximal Chance Ranking Criterion ... 131
  - 5.4.4 Hurwicz Ranking Criterion ... 132
- 5.5 Uncertain Congestion ... 135
- References ... 136

6 Hybrid DEA 139
- 6.1 Symbols and Notations 139
- 6.2 Hybrid DEA Models 140
- 6.3 Hybrid DEA Ranking Criteria 142
  - 6.3.1 Expected Ranking Criterion 142
  - 6.3.2 Optimistic Ranking Criterion 143
  - 6.3.3 Maximal Chance Ranking Criterion 144
  - 6.3.4 Hurwicz Ranking Criterion 145
- 6.4 Hybrid Congestion 146
- References 146

List of Frequently Used Symbols 149

Chapter 1
Uncertain Theories

As real-life decisions are usually made in uncertainty, the motivation is provided for studying the behavior of uncertain phenomena. Many theories have been proposed to handle uncertainty. This chapter mainly introduces four theories crucial for the following chapters, namely, probability theory, credibility theory, uncertainty theory, and uncertain random theory.

1.1 Probability Theory

Probability theory is a branch of mathematics for studying the behavior of random phenomena. The emphasis in this section is mainly on probability space, random variable, probability distribution, independence, operational law, expected value, and variance. The main results in this section are well known. For this reason, the credit references are not provided.

1.1.1 Probability Measure

Let Ω be a nonempty set and $\mathcal{A}$ a σ-algebra over Ω. Each element in $\mathcal{A}$ is called an event. In order to present an axiomatic definition of probability, it is necessary to assign to each event A a number $\Pr\{A\}$ which indicates the probability that A will occur. In order to ensure that the number $\Pr\{A\}$ has certain mathematical properties which we intuitively expect a probability to have, the following three axioms must be satisfied:

Axiom 1.1 (Normality Axiom). $\Pr\{\Omega\} = 1$ *for the universal set* Ω.

Axiom 1.2 (Nonnegativity Axiom). $\Pr\{A\} \geq 0$ *for any event* A.

Axiom 1.3 (Additivity Axiom). *For every countable sequence of mutually disjoint events* $\{A_i\}$, *we have*

$$\Pr\left\{\bigcup_{i=1}^{\infty} A_i\right\} = \sum_{i=1}^{\infty} \Pr\{A_i\}. \qquad (1.1)$$

Definition 1.1. The set function Pr is called a probability measure if it satisfies the normality, nonnegativity, and additivity axioms.

Definition 1.2. Let Ω be a nonempty set, $\mathcal{A}$ a σ-algebra over Ω, and Pr a probability measure. Then the triplet $(\Omega, \mathcal{A}, \Pr)$ is called a probability space.

Theorem 1.1. *Let* $(\Omega, \mathcal{A}, \Pr)$ *be a probability space. Then we have:*

(a) $\Pr\{\emptyset\} = 0$.
(b) $\Pr\{A\} + \Pr\{A^c\} = 1$ *for any* $A \in \mathcal{A}$.
(c) $\Pr\{A\} \leq \Pr\{B\}$ *whenever* $A \subset B$.

Proof. (a) Since $\emptyset$ and Ω are disjoint events and $\emptyset \cup \Omega = \Omega$, we have $\Pr\{\emptyset\} + \Pr\{\Omega\} = \Pr\{\Omega\}$ which makes $\Pr\{\emptyset\} = 0$.
(b) Since A and A^c are disjoint events and $A \cup A^c = \Omega$, we have $\Pr\{A\} + \Pr\{A^c\} = \Pr\{\Omega\} = 1$.
(c) Since $A \subset B$, we have $B = A \cup (B \cap A^c)$, where A and $B \cap A^c$ are disjoint events. Therefore, $\Pr\{B\} = \Pr\{A\} + \Pr\{B \cap A^c\} \geq \Pr\{A\}$.

1.1.2 Random Variable

Definition 1.3. A random variable is a measurable function from a probability space $(\Omega, \mathcal{A}, \Pr)$ to the set of real numbers, i.e., for any Borel set B of real numbers, the set

$$\{\xi \in B\} = \{\omega \in \Omega \mid \xi(\omega) \in B\} \qquad (1.2)$$

is an event.

Definition 1.4. A random variable ξ is said to be:

(a) Nonnegative if $\Pr\{\xi < 0\} = 0$
(b) Positive if $\Pr\{\xi \leq 0\} = 0$
(c) Continuous if $\Pr\{\xi = x\} = 0$ for each $x \in \Re$
(d) Simple if there exists a finite sequence $\{x_1, x_2, \cdots, x_m\}$ such that

$$\Pr\{\xi \neq x_1, \xi \neq x_2, \cdots, \xi \neq x_m\} = 0 \qquad (1.3)$$

(e) Discrete if there exists a countable sequence $\{x_1, x_2, \cdots\}$ such that

$$\Pr\{\xi \neq x_1, \xi \neq x_2, \cdots\} = 0 \qquad (1.4)$$

1.1 Probability Theory

Definition 1.5. The probability distribution $\Phi: \Re \to [0, 1]$ of a random variable ξ is defined by

$$\Phi(x) = \Pr\{\xi \leq x\}. \tag{1.5}$$

That is, $\Phi(x)$ is the probability that the random variable ξ takes a value less than or equal to x.

Theorem 1.2. *A random variable ξ with probability distribution Φ is:*

(a) *Nonnegative if and only if $\Phi(x) = 0$ for all $x < 0$*
(b) *Positive if and only if $\Phi(x) = 0$ for all $x \leq 0$*
(c) *Simple if and only if Φ is a simple function*
(d) *Discrete if and only if Φ is a step function*
(e) *Continuous if and only if Φ is a continuous function*

Proof. Parts (a)–(d) follow immediately from the definition. Next, we prove part (e). If ξ is a continuous random variable, then $\Pr\{\xi = x\} = 0$. It follows from the probability continuity theorem that

$$\lim_{y \uparrow x} (\Phi(x) - \Phi(y)) = \lim_{y \uparrow x} \Pr\{y < \xi \leq x\} = \Pr\{\xi = x\} = 0$$

which proves the left-continuity of Φ. Since a probability distribution is always right-continuous, Φ is continuous. Conversely, if Φ is continuous, then we immediately have $\Pr\{\xi = x\} = 0$ for each $x \in \Re$.

Definition 1.6. The probability density function $\phi: \Re \to [0, +\infty)$ of a random variable ξ is a function such that

$$\Phi(x) = \int_{-\infty}^{x} \phi(y) dy \tag{1.6}$$

holds for all $x \in \Re$, where Φ is the probability distribution of the random variable ξ.

Proof. Let $\mathcal{C}$ be the class of all subsets C of $\Re$ for which the relation

$$\Pr\{\xi \in C\} = \int_C \phi(y) dy \tag{1.7}$$

holds. We will show that $\mathcal{C}$ contains all Borel sets of $\Re$. It follows from the probability continuity theorem and relation (1.7) that $\mathcal{C}$ is a monotone class. It is also clear that $\mathcal{C}$ contains all intervals of the form $(-\infty, a]$, $(a, b]$, (b, ∞) and $\Re$ since

$$\Pr\{\xi \in (-\infty, a]\} = \Phi(a) = \int_{-\infty}^{a} \phi(y) dy,$$

$$\Pr\{\xi \in (b, +\infty)\} = \Phi(+\infty) - \Phi(b) = \int_b^{+\infty} \phi(y)dy,$$

$$\Pr\{\xi \in (a, b]\} = \Phi(b) - \Phi(a) = \int_a^b \phi(y)dy,$$

$$\Pr\{\xi \in \Re\} = \Phi(+\infty) = \int_{-\infty}^{+\infty} \phi(y)dy$$

where Φ is the probability distribution of ξ. Let $\mathcal{F}$ be the class of all finite unions of disjoint sets of the form $(-\infty, a], (a, b], (b, \infty)$ and $\Re$. Note that for any disjoint sets $C_1, C_2, \cdots, C_m$ of $\mathcal{F}$ and $C = C_1 \cup C_2 \cup \cdots \cup C_m$, we have

$$\Pr\{\xi \in C\} = \sum_{j=1}^m \Pr\{\xi \in C_j\} = \sum_{j=1}^m \int_{C_j} \phi(y)dy = \int_C \phi(y)dy.$$

That is, $C \in \mathcal{C}$. Hence, we have $\mathcal{F} \subset \mathcal{C}$. It may also be verified that the class $\mathcal{F}$ is an algebra. Since the smallest σ-algebra containing $\mathcal{F}$ is just the Borel algebra of $\Re$, the monotone class theorem implies that $\mathcal{C}$ contains all Borel sets of $\Re$.

Theorem 1.3 (Probability Inversion Theorem). *Let ξ be a random variable whose probability density function ϕ exists. Then for any Borel set B of $\Re$, we have*

$$\Pr\{\xi \in B\} = \int_B \phi(y)dy. \tag{1.8}$$

1.1.3 Operational Law

Definition 1.7. *The random variables $\xi_1, \xi_2, \cdots, \xi_n$ are said to be independent if*

$$\Pr\left\{\bigcap_{i=1}^n \{\xi_i \in B_i\}\right\} = \prod_{i=1}^n \Pr\{\xi_i \in B_i\} \tag{1.9}$$

for any Borel sets $B_1, B_2, \cdots, B_n$ of real numbers.

Theorem 1.4. *Let ξ_i be random variables with probability distributions Φ_i, $i = 1, 2, \cdots, n$, respectively, and Φ the joint probability distribution of the random vector $(\xi_1, \xi_2, \cdots, \xi_n)$. Then $\xi_1, \xi_2, \cdots, \xi_n$ are independent random variables if and only if*

$$\Phi(x_1, x_2, \cdots, x_n) = \Phi_1(x_1)\Phi_2(x_2)\cdots\Phi_n(x_n) \tag{1.10}$$

for any real numbers $x_1, x_2, \cdots, x_n$.

1.1 Probability Theory

Proof. If $\xi_1, \xi_2, \cdots, \xi_m$ are independent random variables, then we have

$$\Phi(x_1, x_2, \cdots, x_m) = \Pr\{\xi_1 \leq x_1, \xi_2 \leq x_2, \cdots, \xi_m \leq x_m\}$$
$$= \Pr\{\xi_1 \leq x_1\} \Pr\{\xi_2 \leq x_2\} \cdots \Pr\{\xi_m \leq x_m\}$$
$$= \Phi_1(x_1) \Phi_2(x_2) \cdots \Phi_m(x_m)$$

for all $(x_1, x_2, \cdots, x_m) \in \Re^m$.

Conversely, assume that (1.10) holds. Let $x_2, x_3, \cdots, x_m$ be fixed real numbers and $\mathcal{C}$ the class of all subsets C of $\Re$ for which the relation

$$\Pr\{\xi_1 \in C, \xi_2 \leq x_2, \cdots, \xi_m \leq x_m\} = \Pr\{\xi_1 \in C\} \prod_{i=2}^{m} \Pr\{\xi_i \leq x_i\} \quad (1.11)$$

holds. We will show that $\mathcal{C}$ contains all Borel sets of $\Re$. It follows from the probability continuity theorem and relation (1.11) that $\mathcal{C}$ is a monotone class. It is also clear that $\mathcal{C}$ contains all intervals of the form $(-\infty, a], (a, b], (b, \infty)$ and $\Re$. Let $\mathcal{F}$ be the class of all finite unions of disjoint sets of the form $(-\infty, a], (a, b], (b, \infty)$ and $\Re$. Note that for any disjoint sets $C_1, C_2, \cdots, C_k$ of $\mathcal{F}$ and $C = C_1 \cup C_2 \cup \cdots \cup C_k$, we have

$$\Pr\{\xi_1 \in C, \xi_2 \leq x_2, \cdots, \xi_m \leq x_m\}$$
$$= \sum_{j=1}^{m} \Pr\{\xi_1 \in C_j, \xi_2 \leq x_2, \cdots, \xi_m \leq x_m\}$$
$$= \Pr\{\xi_1 \in C\} \Pr\{\xi_2 \leq x_2\} \cdots \Pr\{\xi_m \leq x_m\}.$$

That is, $C \in \mathcal{C}$. Hence, we have $\mathcal{F} \subset \mathcal{C}$. It may also be verified that the class $\mathcal{F}$ is an algebra. Since the smallest σ-algebra containing $\mathcal{F}$ is just the Borel algebra of $\Re$, the monotone class theorem implies that $\mathcal{C}$ contains all Borel sets of $\Re$.

Applying the same reasoning to each ξ_i in turn, we obtain the independence of the random variables.

Theorem 1.5. *Let $\xi_1, \xi_2, \cdots, \xi_n$ be independent random variables and $f_1, f_2, \cdots, f_n$ measurable functions. Then $f_1(\xi_1), f_2(\xi_2), \cdots, f_n(\xi_n)$ are independent random variables.*

Proof. For any Borel sets $B_1, B_2, \cdots, B_n$ of real numbers, it follows from the definition of independence that

$$\Pr\left\{\bigcap_{i=1}^{n}(f_i(\xi_i) \in B_i)\right\} = \Pr\left\{\bigcap_{i=1}^{n}(\xi_i \in f_i^{-1}(B_i))\right\}$$
$$= \prod_{i=1}^{n} \Pr\{\xi_i \in f_i^{-1}(B_i)\} = \prod_{i=1}^{n} \Pr\{f_i(\xi_i) \in B_i\}.$$

Thus $f_1(\xi_1), f_2(\xi_2), \cdots, f_n(\xi_n)$ are independent random variables.

1.1.4 Expected Value

Definition 1.8. Let ξ be a random variable. Then the expected value of ξ is defined by

$$E[\xi] = \int_0^{+\infty} \Pr\{\xi \geq r\} dr - \int_{-\infty}^0 \Pr\{\xi \leq r\} dr \qquad (1.12)$$

provided that at least one of the two integrals is finite.

Example 1.1. Assume that ξ is a discrete random variable taking values x_i with probabilities p_i, $i = 1, 2, \cdots, m$, respectively. It follows from the definition of expected value operator that

$$E[\xi] = \sum_{i=1}^m p_i x_i.$$

Theorem 1.6. *Let ξ be a random variable with probability distribution Φ. If the Lebesgue-Stieltjes integral*

$$\int_{-\infty}^{+\infty} x d\Phi(x)$$

is finite, then we have

$$E[\xi] = \int_{-\infty}^{+\infty} x d\Phi(x). \qquad (1.13)$$

Proof. Since the Lebesgue-Stieltjes integral $\int_{-\infty}^{+\infty} x d\Phi(x)$ is finite, we immediately have

$$\lim_{y \to +\infty} \int_0^y x d\Phi(x) = \int_0^{+\infty} x d\Phi(x), \quad \lim_{y \to -\infty} \int_y^0 x d\Phi(x) = \int_{-\infty}^0 x d\Phi(x)$$

and

$$\lim_{y \to +\infty} \int_y^{+\infty} x d\Phi(x) = 0, \quad \lim_{y \to -\infty} \int_{-\infty}^y x d\Phi(x) = 0.$$

1.1 Probability Theory

It follows from

$$\int_y^{+\infty} x d\Phi(x) \geq y \left(\lim_{z \to +\infty} \Phi(z) - \Phi(y) \right) = y(1 - \Phi(y)) \geq 0, \quad \text{if } y > 0,$$

$$\int_{-\infty}^y x d\Phi(x) \leq y \left(\Phi(y) - \lim_{z \to -\infty} \Phi(z) \right) = y\Phi(y) \leq 0, \quad \text{if } y < 0$$

that

$$\lim_{y \to +\infty} y(1 - \Phi(y)) = 0, \quad \lim_{y \to -\infty} y\Phi(y) = 0.$$

Let $0 = x_0 < x_1 < x_2 < \cdots < x_n = y$ be a partition of $[0, y]$. Then we have

$$\sum_{i=0}^{n-1} x_i \left(\Phi(x_{i+1}) - \Phi(x_i) \right) \to \int_0^y x d\Phi(x)$$

and

$$\sum_{i=0}^{n-1} (1 - \Phi(x_{i+1}))(x_{i+1} - x_i) \to \int_0^y \Pr\{\xi \geq r\} dr$$

as $\max\{|x_{i+1} - x_i| : i = 0, 1, \cdots, n-1\} \to 0$. Since

$$\sum_{i=0}^{n-1} x_i \left(\Phi(x_{i+1}) - \Phi(x_i) \right) - \sum_{i=0}^{n-1} (1 - \Phi(x_{i+1}))(x_{i+1} - x_i) = y(\Phi(y) - 1) \to 0$$

as $y \to +\infty$. This fact implies that

$$\int_0^{+\infty} \Pr\{\xi \geq r\} dr = \int_0^{+\infty} x d\Phi(x).$$

A similar way may prove that

$$-\int_{-\infty}^0 \Pr\{\xi \leq r\} dr = \int_{-\infty}^0 x d\Phi(x).$$

Thus (1.13) is verified by the above two equations.

Theorem 1.7. *Let ξ be a random variable whose expected value exists. Then for any numbers a and b, we have*

$$E[a\xi + b] = aE[\xi] + b. \tag{1.14}$$

Proof. In order to prove the theorem, it suffices to verify that $E[\xi + b] = E[\xi] + b$ and $E[a\xi] = aE[\xi]$. It follows from the expected value operator that, if $b \geq 0$,

$$E[\xi + b] = \int_0^\infty \Pr\{\xi + b \geq r\} dr - \int_{-\infty}^0 \Pr\{\xi + b \leq r\} dr$$

$$= \int_0^\infty \Pr\{\xi \geq r - b\} dr - \int_{-\infty}^0 \Pr\{\xi \leq r - b\} dr$$

$$= E[\xi] + \int_0^b \left(\Pr\{\xi \geq r - b\} + \Pr\{\xi < r - b\}\right) dr$$

$$= E[\xi] + b.$$

If $b < 0$, then we have

$$E[\xi + b] = E[\xi] - \int_b^0 \left(\Pr\{\xi \geq r - b\} + \Pr\{\xi < r - b\}\right) dr = E[\xi] + b.$$

On the other hand, if $a = 0$, then the equation $E[a\xi] = aE[\xi]$ holds trivially. If $a > 0$, we have

$$E[a\xi] = \int_0^\infty \Pr\{a\xi \geq r\} dr - \int_{-\infty}^0 \Pr\{a\xi \leq r\} dr$$

$$= \int_0^\infty \Pr\left\{\xi \geq \frac{r}{a}\right\} dr - \int_{-\infty}^0 \Pr\left\{\xi \leq \frac{r}{a}\right\} dr$$

$$= a\int_0^\infty \Pr\left\{\xi \geq \frac{r}{a}\right\} d\left(\frac{r}{a}\right) - a\int_{-\infty}^0 \Pr\left\{\xi \leq \frac{r}{a}\right\} d\left(\frac{r}{a}\right)$$

$$= aE[\xi].$$

The equation $E[a\xi] = aE[\xi]$ is proved if we verify that $E[-\xi] = -E[\xi]$. In fact,

$$E[-\xi] = \int_0^\infty \Pr\{-\xi \geq r\} dr - \int_{-\infty}^0 \Pr\{-\xi \leq r\} dr$$

$$= \int_0^\infty \Pr\{\xi \leq -r\} dr - \int_{-\infty}^0 \Pr\{\xi \geq -r\} dr$$

$$= \int_{-\infty}^0 \Pr\{\xi \leq r\} dr - \int_0^\infty \Pr\{\xi \geq r\} dr$$

$$= -E[\xi].$$

The proof is finished.

1.1 Probability Theory

Theorem 1.8. *Let ξ and η be random variables with finite expected values. Then we have*

$$E[\xi + \eta] = E[\xi] + E[\eta]. \tag{1.15}$$

Proof. Step 1: We first prove the case where both ξ and η are nonnegative simple random variables taking values $a_1, a_2, \cdots, a_m$ and $b_1, b_2, \cdots, b_n$, respectively. Then $\xi + \eta$ is also a nonnegative simple random variable taking values $a_i + b_j$, $i = 1, 2, \cdots, m$, $j = 1, 2, \cdots, n$. Thus we have

$$E[\xi + \eta] = \sum_{i=1}^{m}\sum_{j=1}^{n}(a_i + b_j)\Pr\{\xi = a_i, \eta = b_j\}$$

$$= \sum_{i=1}^{m}\sum_{j=1}^{n}a_i \Pr\{\xi = a_i, \eta = b_j\} + \sum_{i=1}^{m}\sum_{j=1}^{n}b_j \Pr\{\xi = a_i, \eta = b_j\}$$

$$= \sum_{i=1}^{m}a_i \Pr\{\xi = a_i\} + \sum_{j=1}^{n}b_j \Pr\{\eta = b_j\}$$

$$= E[\xi] + E[\eta].$$

Step 2: Next, we prove the case where ξ and η are nonnegative random variables. For every $i \geq 1$ and every $\omega \in \Omega$, we define

$$\xi_i(\omega) = \begin{cases} \dfrac{k-1}{2^i}, & \text{if } \dfrac{k-1}{2^i} \leq \xi(\omega) < \dfrac{k}{2^i}, \ k = 1, 2, \cdots, i2^i \\ i, & \text{if } i \leq \xi(\omega), \end{cases}$$

$$\eta_i(\omega) = \begin{cases} \dfrac{k-1}{2^i}, & \text{if } \dfrac{k-1}{2^i} \leq \eta(\omega) < \dfrac{k}{2^i}, \ k = 1, 2, \cdots, i2^i \\ i, & \text{if } i \leq \eta(\omega). \end{cases}$$

Then $\{\xi_i\}$, $\{\eta_i\}$, and $\{\xi_i + \eta_i\}$ are three sequences of nonnegative simple random variables such that $\xi_i \uparrow \xi$, $\eta_i \uparrow \eta$ and $\xi_i + \eta_i \uparrow \xi + \eta$ as $i \to \infty$. Note that the functions $\Pr\{\xi_i > r\}$, $\Pr\{\eta_i > r\}$, $\Pr\{\xi_i + \eta_i > r\}$, $i = 1, 2, \cdots$ are also simple. Then we can get

$$\Pr\{\xi_i > r\} \uparrow \Pr\{\xi > r\}, \ \forall r \geq 0$$

as $i \to \infty$. Since the expected value $E[\xi]$ exists, we have

$$E[\xi_i] = \int_0^{+\infty} \Pr\{\xi_i > r\}dr \to \int_0^{+\infty} \Pr\{\xi > r\}dr = E[\xi]$$

as $i \to \infty$. Similarly, we may prove that $E[\eta_i] \to E[\eta]$ and $E[\xi_i + \eta_i] \to E[\xi + \eta]$ as $i \to \infty$. Therefore, $E[\xi + \eta] = E[\xi] + E[\eta]$ since we have proved that $E[\xi_i + \eta_i] = E[\xi_i] + E[\eta_i]$ for $i = 1, 2, \cdots$

Step 3: Finally, if ξ and η are arbitrary random variables, then we define

$$\xi_i(\omega) = \begin{cases} \xi(\omega), & \text{if } \xi(\omega) \geq -i \\ -i, & \text{otherwise,} \end{cases} \qquad \eta_i(\omega) = \begin{cases} \eta(\omega), & \text{if } \eta(\omega) \geq -i \\ -i, & \text{otherwise.} \end{cases}$$

Since the expected values $E[\xi]$ and $E[\eta]$ are finite, we have

$$\lim_{i \to \infty} E[\xi_i] = E[\xi], \quad \lim_{i \to \infty} E[\eta_i] = E[\eta], \quad \lim_{i \to \infty} E[\xi_i + \eta_i] = E[\xi + \eta].$$

Note that $(\xi_i + i)$ and $(\eta_i + i)$ are nonnegative random variables. It follows from Theorem 1.7 that

$$\begin{aligned} E[\xi + \eta] &= \lim_{i \to \infty} E[\xi_i + \eta_i] \\ &= \lim_{i \to \infty} (E[(\xi_i + i) + (\eta_i + i)] - 2i) \\ &= \lim_{i \to \infty} (E[\xi_i + i] + E[\eta_i + i] - 2i) \\ &= \lim_{i \to \infty} (E[\xi_i] + i + E[\eta_i] + i - 2i) \\ &= \lim_{i \to \infty} E[\xi_i] + \lim_{i \to \infty} E[\eta_i] \\ &= E[\xi] + E[\eta] \end{aligned}$$

which proves the theorem.

Theorem 1.9. *Let ξ and η be random variables with finite expected values. Then for any numbers a and b, we have*

$$E[a\xi + b\eta] = aE[\xi] + bE[\eta]. \tag{1.16}$$

Proof. The theorem follows immediately from Theorems 1.7 and 1.8.

Definition 1.9. *Let ξ be a random variable with finite expected value e. Then the variance of ξ is defined by $V[\xi] = E[(\xi - e)^2]$.*

The variance of a random variable provides a measure of the spread of the distribution around its expected value. A small value of variance indicates that the random variable is tightly concentrated around its expected value; and a large value of variance indicates that the random variable has a wide spread around its expected value.

Theorem 1.10. *If ξ is a random variable whose variance exists, a and b are real numbers, then $V[a\xi + b] = a^2 V[\xi]$.*

Proof. It follows from the definition of variance that

$$V[a\xi + b] = E\left[(a\xi + b - aE[\xi] - b)^2\right] = a^2 E[(\xi - E[\xi])^2] = a^2 V[\xi].$$

1.2 Credibility Theory

The concept of fuzzy set was initiated by Zadeh [23] via membership function in 1965. In order to measure a fuzzy event, Zadeh [24] proposed the concept of possibility measure. Although possibility measure has been widely used, it has no self-duality property. However, a self-dual measure is absolutely needed in both theory and practice. In order to define a self-dual measure, Liu and Liu [19] presented the concept of credibility measure. In addition, a sufficient and necessary condition for credibility measure was given by Li and Liu [4]. Credibility theory, founded by Liu [6] in 2004 and refined by Liu [7] in 2007, is a branch of mathematics for studying the behavior of fuzzy phenomena.

The emphasis in this section is mainly on credibility measure, credibility space, fuzzy variable, membership function, credibility distribution, independence, expected value, variance, and so on.

1.2.1 Credibility Measure

Let Θ be a nonempty set and $\mathcal{P}(\Theta)$ the power set of Θ. For any $A \in \mathcal{P}(\Theta)$, we use a credibility measure $\text{Cr}\{A\}$ to express the chance that fuzzy event A occurs. In order to ensure that the number $\text{Cr}\{A\}$ has certain mathematical properties which we intuitively expect a credibility to have:

Axiom 1.4 (Normality). $\text{Cr}\{\Theta\} = 1$.

Axiom 1.5 (Monotonicity). $\text{Cr}\{A\} \leq \text{Cr}\{B\}$ *whenever* $A \subset B$.

Axiom 1.6 (Self-Duality). $\text{Cr}\{A\} + \text{Cr}\{A^c\} = 1$ *for any event A*.

Axiom 1.7 (Maximality). $\text{Cr}\{\cup_i A_i\} = \sup_i \text{Cr}\{A_i\}$ *for any events* $\{A_i\}$ *with* $\sup_i \text{Cr}\{A_i\} < 0.5$.

Definition 1.10 (Liu and Liu [19]). The set function Cr is called a credibility measure if it satisfies the normality, monotonicity, self-duality, and maximality axioms.

Theorem 1.11 (Liu [6], Credibility Subadditivity Theorem). *The credibility measure is subadditive. That is,*

$$\mathrm{Cr}\{A \cup B\} \leq \mathrm{Cr}\{A\} + \mathrm{Cr}\{B\} \quad (1.17)$$

for any A and B.

Proof. The argument breaks down into three cases.

Case 1: $\mathrm{Cr}\{A\} < 0.5$ and $\mathrm{Cr}\{B\} < 0.5$. It follows from Axiom 1.7 that

$$\mathrm{Cr}\{A \cup B\} = \mathrm{Cr}\{A\} \vee \mathrm{Cr}\{B\} \leq \mathrm{Cr}\{A\} + \mathrm{Cr}\{B\}.$$

Case 2: $\mathrm{Cr}\{A\} \geq 0.5$. For this case, by using Axioms 1.5 and 1.6, we have $\mathrm{Cr}\{A^c\} \leq 0.5$ and $\mathrm{Cr}\{A \cup B\} \geq \mathrm{Cr}\{A\} \geq 0.5$. Then

$$\mathrm{Cr}\{A^c\} = \mathrm{Cr}\{A^c \cap B\} \vee \mathrm{Cr}\{A^c \cap B^c\} \quad (1.18)$$

$$\leq \mathrm{Cr}\{A^c \cap B\} + \mathrm{Cr}\{A^c \cap B^c\} \quad (1.19)$$

$$\leq \mathrm{Cr}\{B\} + \mathrm{Cr}\{A^c \cap B^c\}. \quad (1.20)$$

Applying this inequality, we obtain

$$\mathrm{Cr}\{A\} + \mathrm{Cr}\{B\} = 1 - \mathrm{Cr}\{A^c\} + \mathrm{Cr}\{B\} \quad (1.21)$$

$$\geq 1 - \mathrm{Cr}\{B\} - \mathrm{Cr}\{A^c \cap B^c\} + \mathrm{Cr}\{B\} \quad (1.22)$$

$$= 1 - \mathrm{Cr}\{A^c \cap B^c\} \quad (1.23)$$

$$= \mathrm{Cr}\{A \cup B\}. \quad (1.24)$$

Case 3: $\mathrm{Cr}\{B\} \geq 0.5$. This case may be proved by a similar process of Case 2. The theorem is proved.

Theorem 1.12 (Liu [6], Credibility Semicontinuity Law). *For any event $A_1, A_2, \cdots$, we have*

$$\lim_{i \to \infty} \mathrm{Cr}\{A_i\} = \mathrm{Cr}\left\{\lim_{i \to \infty} A_i\right\} \quad (1.25)$$

if one of the following conditions is satisfied:

(a) $\mathrm{Cr}\{A\} \leq 0.5$ and $A_i \uparrow A$. (b) $\lim_{i \to \infty} \mathrm{Cr}\{A_i\} < 0.5$ and $A_i \uparrow A$.
(c) $\mathrm{Cr}\{A\} \geq 0.5$ and $A_i \downarrow A$. (d) $\lim_{i \to \infty} \mathrm{Cr}\{A_i\} > 0.5$ and $A_i \downarrow A$.

Proof. (a) Since $\mathrm{Cr}\{A\} \leq 0.5$, we have $\mathrm{Cr}\{A_i\} \leq 0.5$ for each i. It follows from Axiom 1.7 that

$$\mathrm{Cr}\{A\} = \mathrm{Cr}\{\cup_i A_i\} = \sup_i \mathrm{Cr}\{A_i\} = \lim_{i \to \infty} \mathrm{Cr}\{A_i\}.$$

(b) Since $\lim_{i\to\infty} \text{Cr}\{A_i\} < 0.5$, we have $\sup_i \text{Cr}\{A_i\} < 0.5$. It follows from Axiom 1.7 that

$$\text{Cr}\{A\} = \text{Cr}\{\cup_i A_i\} = \sup_i \text{Cr}\{A_i\} = \lim_{i\to\infty} \text{Cr}\{A_i\}.$$

(c) Since $\text{Cr}\{A\} \geq 0.5$ and $A_i \downarrow A$, it follows from the self-duality of credibility measure that $\text{Cr}\{A^c\} \leq 0.5$ and $A_i^c \uparrow A^c$. Thus $\text{Cr}\{A_i\} = 1 - \text{Cr}\{A_i^c\} \to 1 - \text{Cr}\{A^c\} = \text{Cr}\{A\}$ as $i \to \infty$.

(d) Since $\lim_{i\to\infty} \text{Cr}\{A_i\} > 0.5$ and $A_i \downarrow A$, it follows from the self-duality of credibility measure that

$$\lim_{i\to\infty} \text{Cr}\{A_i^c\} = \lim_{i\to\infty} (1 - \text{Cr}\{A_i\}) < 0.5$$

and $A_i^c \uparrow A^c$. Thus $\text{Cr}\{A_i\} = 1 - \text{Cr}\{A_i^c\} \to 1 - \text{Cr}\{A^c\} = \text{Cr}\{A\}$ as $i \to \infty$. The theorem is proved.

Theorem 1.13 (Credibility Asymptotic Law). *For any event $A_1, A_2, \cdots$, we have*

$$\lim_{i\to\infty} \text{Cr}\{A_i\} \geq 0.5, \quad \text{if } A_i \uparrow \Theta, \tag{1.26}$$

$$\lim_{i\to\infty} \text{Cr}\{A_i\} \leq 0.5, \quad \text{if } A_i \downarrow \emptyset. \tag{1.27}$$

Proof. Assume $A_i \uparrow \Theta$. If $\lim_{i\to\infty} \text{Cr}\{A_i\} < 0.5$, it follows from the credibility semicontinuity law that

$$\text{Cr}\{\Theta\} = \lim_{i\to\infty} \text{Cr}\{A_i\} < 0.5$$

which is in contradiction with $\text{Cr}\{\Theta\} = 1$. The first inequality is proved. The second one may be verified similarly.

Theorem 1.14 (Li and Liu [4], Credibility Extension Theorem). *Suppose that Θ is a nonempty set and $\text{Cr}\{\theta\}$ is a nonnegative function on Θ satisfying the credibility extension condition,*

$$\begin{array}{c} \sup_{\theta \in \Theta} \text{Cr}\{\theta\} \geq 0.5, \\ \text{Cr}\{\theta^*\} + \sup_{\theta \neq \theta^*} \text{Cr}\{\theta\} = 1 \text{ if } \text{Cr}\{\theta^*\} \geq 0.5. \end{array} \tag{1.28}$$

Then $\text{Cr}\{\theta\}$ has a unique extension to a credibility measure as follows:

$$\text{Cr}\{A\} = \begin{cases} \sup_{\theta \in A} \text{Cr}\{\theta\}, & \text{if } \sup_{\theta \in A} \text{Cr}\{\theta\} < 0.5 \\ 1 - \sup_{\theta \in A^c} \text{Cr}\{\theta\}, & \text{if } \sup_{\theta \in A} \text{Cr}\{\theta\} \geq 0.5. \end{cases} \tag{1.29}$$

Proof. We first prove that the set function $\mathrm{Cr}\{A\}$ defined by (1.29) is a credibility measure.

Step 1: By the credibility extension condition $\sup_{\theta \in \Theta} \mathrm{Cr}\{\theta\} \geq 0.5$, we have

$$\mathrm{Cr}\{\Theta\} = 1 - \sup_{\theta \in \emptyset} \mathrm{Cr}\{\theta\} = 1 - 0 = 1.$$

Step 2: If $A \subset B$, then $B^c \subset A^c$. The proof breaks down into two cases.

Case 1: $\sup_{\theta \in A} \mathrm{Cr}\{\theta\} < 0.5$. For this case, we have

$$\mathrm{Cr}\{A\} = \sup_{\theta \in A} \mathrm{Cr}\{\theta\} \leq \sup_{\theta \in B} \mathrm{Cr}\{\theta\} \leq \mathrm{Cr}\{B\}.$$

Case 2: $\sup_{\theta \in A} \mathrm{Cr}\{\theta\} \geq 0.5$. For this case, we have $\sup_{\theta \in B} \mathrm{Cr}\{\theta\} \geq 0.5$. Thus,

$$\mathrm{Cr}\{A\} = 1 - \sup_{\theta \in A^c} \mathrm{Cr}\{\theta\} \leq 1 - \sup_{\theta \in B^c} \mathrm{Cr}\{\theta\} = \mathrm{Cr}\{B\}.$$

Step 3: In order to prove $\mathrm{Cr}\{A\} + \mathrm{Cr}\{A^c\} = 1$, the argument breaks down into two cases.

Case 1: $\sup_{\theta \in A} \mathrm{Cr}\{\theta\} < 0.5$. For this case, we have $\sup_{\theta \in A^c} \mathrm{Cr}\{\theta\} \geq 0.5$. Thus

$$\mathrm{Cr}\{A\} + \mathrm{Cr}\{A^c\} = \sup_{\theta \in A} \mathrm{Cr}\{\theta\} + 1 - \sup_{\theta \in A} \mathrm{Cr}\{\theta\} = 1.$$

Case 2: $\sup_{\theta \in A} \mathrm{Cr}\{\theta\} \geq 0.5$. For this case, we have $\sup_{\theta \in A^c} \mathrm{Cr}\{\theta\} \leq 0.5$, and

$$\mathrm{Cr}\{A\} + \mathrm{Cr}\{A^c\} = 1 - \sup_{\theta \in A^c} \mathrm{Cr}\{\theta\} + \sup_{\theta \in A^c} \mathrm{Cr}\{\theta\} = 1.$$

Step 4: For any collection $\{A_i\}$ with $\sup_i \mathrm{Cr}\{A_i\} < 0.5$, we have

$$\mathrm{Cr}\{\cup_i A_i\} = \sup_{\theta \in \cup_i A_i} \mathrm{Cr}\{\theta\} = \sup_i \sup_{\theta \in A_i} \mathrm{Cr}\{\theta\} = \sup_i \mathrm{Cr}\{A_i\}.$$

Thus Cr is a credibility measure because it satisfies the four axioms.

Finally, let us prove the uniqueness. Assume that Cr_1 and Cr_2 are two credibility measures such that $\mathrm{Cr}_1\{\theta\} = \mathrm{Cr}_2\{\theta\}$ for each $\theta \in \Theta$. Let us prove that $\mathrm{Cr}_1\{A\} = \mathrm{Cr}_2\{A\}$ for any event A. The argument breaks down into three cases.

Case 1: $\mathrm{Cr}_1\{A\} < 0.5$. For this case, it follows from Axiom 1.7 that

$$\mathrm{Cr}_1\{A\} = \sup_{\theta \in A} \mathrm{Cr}_1\{\theta\} = \sup_{\theta \in A} \mathrm{Cr}_2\{\theta\} = \mathrm{Cr}_2\{A\}.$$

Case 2: $Cr_1\{A\} > 0.5$. For this case, we have $Cr_1\{A^c\} < 0.5$. It follows from the first case that $Cr_1\{A^c\} = Cr_2\{A^c\}$ which implies $Cr_1\{A\} = Cr_2\{A\}$.
Case 3: $Cr_1\{A\} = 0.5$. For this case, we have $Cr_1\{A^c\} = 0.5$, and

$$Cr_2\{A\} \geq \sup_{\theta \in A} Cr_2\{\theta\} = \sup_{\theta \in A} Cr_1\{\theta\} = Cr_1\{A\} = 0.5,$$

$$Cr_2\{A^c\} \geq \sup_{\theta \in A^c} Cr_2\{\theta\} = \sup_{\theta \in A^c} Cr_1\{\theta\} = Cr_1\{A^c\} = 0.5.$$

Hence, $Cr_2\{A\} = 0.5 = Cr_1\{A\}$. The uniqueness is proved.

Definition 1.11. Let Θ be a nonempty set, $\mathcal{P}$ the power set of Θ, and Cr a credibility measure. Then the triplet $(\Theta, \mathcal{P}, Cr)$ is called a credibility space.

Axiom 1.8 (Product Credibility Axiom). *Let Θ_k be nonempty sets on which Cr_k are credibility measures, $k = 1, 2, \cdots, n$, respectively, and $\Theta = \Theta_1 \times \Theta_2 \times \cdots \times \Theta_n$. Then $Cr\{(\theta_1, \theta_2, \cdots, \theta_n)\} = Cr_1\{\theta_1\} \wedge Cr_2\{\theta_2\} \wedge \cdots \wedge Cr_n\{\theta_n\}$ for each $(\theta_1, \theta_2, \cdots, \theta_n) \in \Theta$.*

Theorem 1.15 (Product Credibility Theory). *Suppose that Θ_k are nonempty sets, Cr_k the credibility measures on $\mathcal{P}_k$, $k = 1, 2, \cdots, n$, respectively. Let $\Theta = \Theta_1 \times \Theta_2 \times \cdots \times \Theta_n$. Then $Cr = Cr_1 \wedge Cr_2 \wedge \cdots \wedge Cr_n$ defined by Axiom 1.8 has a unique extension to a credibility measure on $\mathcal{P}$ as follows:*

$$Cr\{A\} = \begin{cases} \sup_{(\theta_1,\theta_2,\cdots,\theta_n) \in A} \min_{1 \leq k \leq n} Cr_k\{\theta_k\}, \\ \quad \text{if } \sup_{(\theta_1,\theta_2,\cdots,\theta_n) \in A} \min_{1 \leq k \leq n} Cr_k\{\theta_k\} < 0.5 \\ 1 - \sup_{(\theta_1,\theta_2,\cdots,\theta_n) \in A^c} \min_{1 \leq k \leq n} Cr_k\{\theta_k\}, \\ \quad \text{if } \sup_{(\theta_1,\theta_2,\cdots,\theta_n) \in A} \min_{1 \leq k \leq n} Cr_k\{\theta_k\} \geq 0.5 \end{cases}$$

for each $A \in \mathcal{P}$.

Proof. For each $\theta = (\theta_1, \theta_2, \cdots, \theta_n) \in \Theta$, we have $Cr\{\theta\} = Cr_1\{\theta_1\} \wedge Cr_2\{\theta_2\} \wedge \cdots \wedge Cr_n\{\theta_n\}$. Let us prove that $Cr\{\theta\}$ satisfies the credibility extension condition. Since $\sup_{\theta_k \in \Theta_k} Cr\{\theta_k\} \geq 0.5$ for each k, we have

$$\sup_{\theta \in \Theta} Cr\{\theta\} = \sup_{(\theta_1,\theta_2,\cdots,\theta_n) \in \Theta} \min_{1 \leq k \leq n} Cr_k\{\theta_k\} \geq 0.5.$$

Now, we suppose that $\theta^* = (\theta_1^*, \theta_2^*, \cdots, \theta_n^*)$ is a point with $Cr\{\theta^*\} \geq 0.5$. Without loss of generality, let i be the index such that

$$Cr\{\theta^*\} = \min_{1 \leq k \leq n} Cr_k\{\theta_k^*\} = Cr_i\{\theta_i^*\}. \tag{1.30}$$

We also immediately have

$$\mathrm{Cr}_k\{\theta_k^*\} \geq 0.5, k = 1, 2, \cdots n; \quad (1.31)$$

$$\mathrm{Cr}_k\{\theta_k^*\} + \sup_{\theta_k \neq \theta_k^*} \mathrm{Cr}_k\{\theta_k\} = 1, k = 1, 2, \cdots, n; \quad (1.32)$$

$$\sup_{\theta_i \neq \theta_i^*} \mathrm{Cr}_i\{\theta_i^*\} \geq \sup_{\theta_k \neq \theta_k^*} \mathrm{Cr}_k\{\theta_k\}, k = 1, 2, \cdots, n; \quad (1.33)$$

$$\sup_{\theta_k \neq \theta_k^*} \mathrm{Cr}_k\{\theta_k\} \leq 0.5, k = 1, 2, \cdots, n. \quad (1.34)$$

It follows from (1.31) and (1.34) that

$$\sup_{\theta \neq \theta^*} \mathrm{Cr}\{\theta\} = \sup_{(\theta_1,\theta_2,\cdots,\theta_n) \neq (\theta_1^*,\theta_2^*,\cdots,\theta_n^*)} \min_{1 \leq k \leq n} \mathrm{Cr}_k\{\theta_k\}$$

$$\geq \sup_{\theta_i \neq \theta_i^*} \min_{1 \leq k \leq i-1} \mathrm{Cr}_k\{\theta_k^*\} \wedge \mathrm{Cr}_i\{\theta_i\} \wedge \min_{i+1 \leq k \leq n} \mathrm{Cr}_k\{\theta_k^*\}$$

$$= \sup_{\theta_i \neq \theta_i^*} \mathrm{Cr}_i\{\theta_i\}.$$

We next suppose that

$$\sup_{\theta \neq \theta^*} \mathrm{Cr}\{\theta\} > \sup_{\theta_i \neq \theta_i^*} \mathrm{Cr}_i\{\theta_i\}.$$

Then there is a point $(\theta_1', \theta_2', \cdots, \theta_n') \neq (\theta_1^*, \theta_2^*, \cdots, \theta_n^*)$ such that

$$\min_{1 \leq k \leq n} \mathrm{Cr}_k\{\theta_k'\} > \sup_{\theta_i \neq \theta_i^*} \mathrm{Cr}_i\{\theta_i\}.$$

Let j be one of the index such that $\theta_j' \neq \theta_j^*$. Then

$$\mathrm{Cr}_j\{\theta_j'\} > \sup_{\theta_i \neq \theta_i^*} \mathrm{Cr}_i\{\theta_i\}.$$

That is,

$$\sup_{\theta_j \neq \theta_j^*} \mathrm{Cr}_j\{\theta_j\} > \sup_{\theta_i \neq \theta_i^*} \mathrm{Cr}_i\{\theta_i\}$$

which is in contradiction with (1.33). Thus

$$\sup_{\theta \neq \theta^*} \mathrm{Cr}\{\theta\} = \sup_{\theta_i \neq \theta_i^*} \mathrm{Cr}_i\{\theta_i\}. \quad (1.35)$$

It follows from (1.30), (1.32), and (1.35) that

$$\mathrm{Cr}\{\theta^*\} + \sup_{\theta \neq \theta^*} \mathrm{Cr}\{\theta\} = \mathrm{Cr}_i\{\theta_i^*\} + \sup_{\theta_i \neq \theta_i^*} \mathrm{Cr}_i\{\theta_i\} = 1.$$

Thus Cr satisfies the credibility extension condition. It follows from the credibility extension theorem that $\mathrm{Cr}\{A\}$ is just the unique extension of $\mathrm{Cr}\{\theta\}$. The theorem is proved.

Definition 1.12. Let $(\Theta_k, \mathcal{P}_k, \mathrm{Cr}_k), k = 1, 2, \cdots, n$ be credibility spaces, $\Theta = \Theta_1 \times \Theta_2 \times \cdots \times \Theta_n$ and $\mathrm{Cr} = \mathrm{Cr}_1 \wedge \mathrm{Cr}_2 \wedge \cdots \wedge \mathrm{Cr}_n$. Then $(\Theta, \mathcal{P}, \mathrm{Cr})$ is called the product credibility space.

1.2.2 Fuzzy Variable

Definition 1.13. A fuzzy variable is a (measurable) function from a credibility space $(\Theta, \mathcal{P}, \mathrm{Cr})$ to the set of real numbers.

Definition 1.14. Let ξ be a fuzzy variable defined on the credibility space $(\Theta, \mathcal{P}(\Theta), \mathrm{Cr})$. Then its membership function is derived from the credibility measure by

$$\mu(x) = (2\mathrm{Cr}\{\xi = x\}) \wedge 1, \quad x \in \Re. \tag{1.36}$$

Theorem 1.16 (Sufficient and Necessary Condition for Membership Function). *A function $\mu : \Re \to [0, 1]$ is a membership function if and only if $\sup \mu(x) = 1$.*

Proof. If μ is a membership function, then there exists a fuzzy variable ξ whose membership function is just μ, and

$$\sup_{x \in \Re} \mu(x) = \sup_{x \in \Re}(2\mathrm{Cr}\{\xi = x\}) \wedge 1.$$

If there is some point $x \in \Re$ such that $\mathrm{Cr}\{\xi = x\} \geq 0.5$, then $\sup \mu(x) = 1$. Otherwise, we have $\mathrm{Cr}\{\xi = x\} < 0.5$ for each $x \in \Re$. It follows from Axiom 1.7 that

$$\sup_{x \in \Re} \mu(x) = \sup_{x \in \Re}(2\mathrm{Cr}\{\xi = x\}) \wedge 1 = 2 \sup_{x \in \Re} \mathrm{Cr}\{\xi = x\} = 2(\mathrm{Cr}\{\Theta\} \wedge 0.5) = 1.$$

Conversely, suppose that $\sup \mu(x) = 1$. For each $x \in \Re$, we define

$$\mathrm{Cr}\{x\} = \frac{1}{2}\left(\mu(x) + 1 - \sup_{y \neq x} \mu(y)\right).$$

It is clear that

$$\sup_{x \in \Re} \text{Cr}\{x\} \geq \frac{1}{2}(1 + 1 - 1) = 0.5.$$

For any $x^* \in \Re$ with $\text{Cr}\{x^*\} \geq 0.5$, we have $\mu(x^*) = 1$ and

$$\text{Cr}\{x^*\} + \sup_{y \neq x^*} \text{Cr}\{y\}$$

$$= \frac{1}{2}\left(\mu(x^*) + 1 - \sup_{y \neq x^*} \mu(y)\right) + \sup_{y \neq x^*} \frac{1}{2}\left(\mu(y) + 1 - \sup_{z \neq y} \mu(z)\right)$$

$$= 1 - \frac{1}{2}\sup_{y \neq x^*} \mu(y) + \frac{1}{2}\sup_{y \neq x^*} \mu(y) = 1.$$

Thus $\text{Cr}\{x\}$ satisfies the credibility extension condition and has a unique extension to credibility measure on $P(\Re)$ by using the credibility extension theorem. Now, we define a fuzzy variable ξ as an identity function from the credibility space $(\Re, P(\Re), \text{Cr})$ to $\Re$. Then the membership function of the fuzzy variable ξ is

$$(2\text{Cr}\{\xi = x\}) \wedge 1 = \left(\mu(x) + 1 - \sup_{y \neq x} \mu(y)\right) \wedge 1 = \mu(x)$$

for each x. The theorem is proved.

By an equipossible fuzzy variable (a, b) we mean the fuzzy variable whose membership function is $\mu(x) = \begin{cases} 1, & \text{if } a \leq x \leq b \\ 0, & \text{otherwise.} \end{cases}$

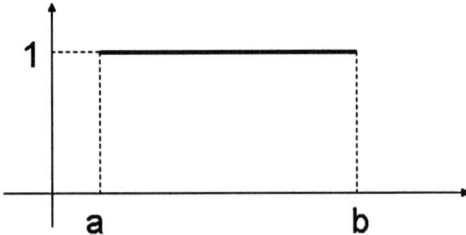

By a triangular fuzzy variable we mean the fuzzy variable fully determined by the triplet (a, b, c) of crisp numbers with $a < b < c$, whose membership function is given by

1.2 Credibility Theory

$$\mu(x) = \begin{cases} \dfrac{x-a}{b-a}, & \text{if } a \le x \le b \\ \dfrac{x-c}{b-c}, & \text{if } b \le x \le c \\ 0, & \text{otherwise.} \end{cases}$$

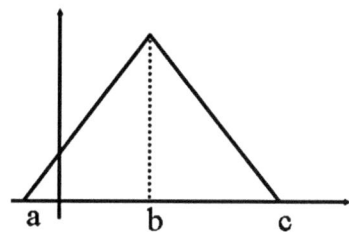

By a trapezoidal fuzzy variable we mean the fuzzy variable fully determined by the quadruplet (a,b,c,d) of crisp numbers with $a < b < c < d$, whose membership function is given by

$$\mu(x) = \begin{cases} \dfrac{x-a}{b-a}, & \text{if } a \le x \le b \\ 1, & \text{if } b \le x \le c \\ \dfrac{x-d}{c-d}, & \text{if } c \le x \le d \\ 0, & \text{otherwise.} \end{cases}$$

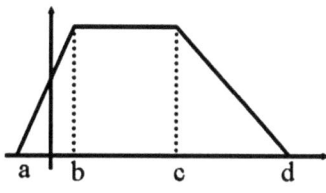

Theorem 1.17 (Credibility Inversion Theorem). *Let ξ be a fuzzy variable with membership function μ. Then for any set B of real numbers, we have*

$$\mathrm{Cr}\{\xi \in B\} = \frac{1}{2}\left(\sup_{x \in B} \mu(x) + 1 - \sup_{x \in B^c} \mu(x)\right). \tag{1.37}$$

Proof. If $\mathrm{Cr}\{\xi \in B\} \le 0.5$, then by Axiom 1.5, we have $\mathrm{Cr}\{\xi \in x\} \le 0.5$ for each $x \in B$. It follows from Axiom 1.7 that

$$\mathrm{Cr}\{\xi \in B\} = \frac{1}{2}\left(\sup_{x \in B} (2\mathrm{Cr}\{\xi = x\} \wedge 1)\right) = \frac{1}{2}\sup_{x \in B} \mu(x). \tag{1.38}$$

The self-duality of credibility measure implies that $\mathrm{Cr}\{\xi \in B^c\} \ge 0.5$ and $\sup_{x \in B^c}\mathrm{Cr}\{\xi = x\} \ge 0.5$, i.e.,

$$\sup_{x \in B^c} \mu(x) = \sup_{x \in B^c} (2\mathrm{Cr}\{\xi = x\} \wedge 1) = 1. \tag{1.39}$$

It follows from (1.38) and (1.39) that (1.37) holds.

If $\mathrm{Cr}\{\xi \in B\} \ge 0.5$, then $\mathrm{Cr}\{\xi \in B^c\} \le 0.5$. It follows from the first case that

$$\mathrm{Cr}\{\xi \in B\} = 1 - \mathrm{Cr}\{\xi \in B^c\} = 1 - \frac{1}{2}\left(\sup_{x \in B^c} \mu(x) + 1 - \sup_{x \in B} \mu(x)\right)$$

$$= \frac{1}{2}\left(\sup_{x \in B} \mu(x) + 1 - \sup_{x \in B^c} \mu(x)\right).$$

The theorem is proved.

Definition 1.15 (Liu [5]). The credibility distribution $\Phi : \Re \to [0, 1]$ of a fuzzy variable ξ is defined by

$$\Phi(x) = \text{Cr}\{\theta \in \Theta \mid \xi(\theta) \leq x\}. \tag{1.40}$$

Definition 1.16 ([5]). The credibility density function $\phi: \Re \to [0, +\infty)$ of a fuzzy variable ξ is a function such that

$$\Phi(x) = \int_{-\infty}^{x} \phi(y) dy, \quad \forall x \in \Re, \tag{1.41}$$

$$\int_{-\infty}^{+\infty} \phi(y) dy = 1 \tag{1.42}$$

where Φ is the credibility distribution of the fuzzy variable ξ.

Theorem 1.18. *Let ξ be a fuzzy variable whose credibility density function ϕ exists. Then we have*

$$\text{Cr}\{\xi \leq x\} = \int_{-\infty}^{x} \phi(y) dy, \quad \text{Cr}\{\xi \geq x\} = \int_{x}^{+\infty} \phi(y) dy. \tag{1.43}$$

Proof. The first part follows immediately from the definition. In addition, by the self-duality of credibility measure, we have

$$\text{Cr}\{\xi \geq x\} = 1 - \text{Cr}\{\xi < x\} = \int_{-\infty}^{+\infty} \phi(y) dy - \int_{-\infty}^{x} \phi(y) dy = \int_{x}^{+\infty} \phi(y) dy.$$

The theorem is proved.

Definition 1.17 (Liu and Gao [17]). The fuzzy variables $\xi_1, \xi_2, \cdots, \xi_m$ are said to be independent if and only if

$$\text{Cr}\left\{\bigcap_{i=1}^{m} \{\xi_i \in B_i\}\right\} = \min_{1 \leq i \leq m} \text{Cr}\{\xi_i \in B_i\}$$

for any sets $B_1, B_2, \cdots, B_m$ of $\Re$.

Theorem 1.19. *The fuzzy variables $\xi_1, \xi_2, \cdots, \xi_m$ are independent if and only if*

$$\text{Cr}\left\{\bigcup_{i=1}^{m} \{\xi_i \in B_i\}\right\} = \max_{1 \leq i \leq m} \text{Cr}\{\xi_i \in B_i\}$$

for any sets $B_1, B_2, \cdots, B_m$ of $\Re$.

1.2 Credibility Theory

Proof. It follows from the self-duality of credibility measure that $\xi_1, \xi_2, \ldots, \xi_m$ are independent if and only if

$$\mathrm{Cr}\left\{\bigcup_{i=1}^{m}\{\xi_i \in B_i\}\right\} = 1 - \mathrm{Cr}\left\{\bigcap_{i=1}^{m}\{\xi_i \in B_i^c\}\right\}$$

$$= 1 - \min_{1 \le i \le m} \mathrm{Cr}\{\xi_i \in B_i^c\} = \max_{1 \le i \le m} \mathrm{Cr}\{\xi_i \in B_i\}.$$

Thus (1.19) is verified. The proof is complete.

Theorem 1.20. *Let ξ_i be independent fuzzy variables and $f_i : \Re \to \Re$ functions, $i = 1, 2, \cdots, m$. Then $f_1(\xi_1), f_2(\xi_2), \cdots, f_m(\xi_m)$ are independent fuzzy variables.*

Proof. For any sets $B_1, B_2, \cdots, B_m$ of $\Re$, we have

$$\mathrm{Cr}\left\{\bigcap_{i=1}^{m}\{f_i(\xi_i) \in B_i\}\right\} = \mathrm{Cr}\left\{\bigcap_{i=1}^{m}\{\xi_i \in f_i^{-1}(B_i)\}\right\}$$

$$= \min_{1 \le i \le m} \mathrm{Cr}\{\xi_i \in f_i^{-1}(B_i)\} = \min_{1 \le i \le m} \mathrm{Cr}\{f_i(\xi_i) \in B_i\}.$$

Thus $f_1(\xi_1), f_2(\xi_2), \cdots, f_m(\xi_m)$ are independent fuzzy variables.

Theorem 1.21 (Extension Principle of Zadeh). *Let $\xi_1, \xi_2, \cdots, \xi_n$ be independent fuzzy variables with membership functions $\mu_1, \mu_2, \cdots, \mu_n$, respectively, and $f : \Re^n \to \Re$ a function. Then the membership function μ of $\xi = f(\xi_1, \xi_2, \cdots, \xi_n)$ is derived from the membership functions $\mu_1, \mu_2, \cdots, \mu_n$ by*

$$\mu(x) = \sup_{x = f(x_1, x_2, \cdots, x_n)} \min_{1 \le i \le n} \mu_i(x_i).$$

Proof. It follows from Definition 1.14 that the membership function of $\xi = f(\xi_1, \xi_2, \ldots, \xi_n)$ is

$$mu(x) = (2\mathrm{Cr}\{f(\xi_1, \xi_2, \ldots, \xi_n) = x\}) \wedge 1$$

$$= \left(2\mathrm{Cr}\left\{\bigcup_{x = f(x_1, x_2, \ldots, x_n)} \{\xi_1 = x_1, \xi_2 = x_2, \ldots, \xi_n = x_n\}\right\}\right) \wedge 1$$

$$= \left(2 \sup_{x = f(x_1, x_2, \ldots, x_n)} \mathrm{Cr}\{\xi_2 = x_2, \ldots, \xi_n = x_n\}\right) \wedge 1$$

$$= \left(2 \sup_{x = f(x_1, x_2, \ldots, x_n)} \min_{1 \le k \le n} \mathrm{Cr}\{\xi_i = x_i\}\right) \wedge 1 \quad \text{(by independence)}$$

$$= \sup_{x = f(x_1, x_2, \ldots, x_n)} \min_{1 \le k \le n} (2\mathrm{Cr}\{\xi_i = x_i\}) \wedge 1$$

$$= \sup_{x = f(x_1, x_2, \ldots, x_n)} \min_{1 \le i \le n} \mu_i(x_i)$$

Example 1.2. The sum of independent equipossible fuzzy variables $\xi = (a_1, a_2)$ and $\eta = (b_1, b_2)$ is also an equipossible fuzzy variable,

$$\xi + \eta = (a_1 + b_1, a_2 + b_2).$$

Their product is also an equipossible fuzzy variable,

$$\xi \cdot \eta = \left(\min_{a_1 \leq x \leq a_2, b_1 \leq y \leq b_2} xy, \max_{a_1 \leq x \leq a_2, b_1 \leq y \leq b_2} xy \right).$$

Example 1.3. The sum of independent triangular fuzzy variables $\xi = (a_1, a_2, a_3)$ and $\eta = (b_1, b_2, b_3)$ is also a triangular fuzzy variable,

$$\xi + \eta = (a_1 + b_1, a_2 + b_2, a_3 + b_3).$$

The product of a triangular fuzzy variable $\xi = (a_1, a_2, a_3)$ and a scalar number λ is also a triangular fuzzy variable,

$$\lambda \cdot \xi = \begin{cases} (\lambda a_1, \lambda a_2, \lambda a_3), & \text{if } \lambda \geq 0 \\ (\lambda a_3, \lambda a_2, \lambda a_1), & \text{if } \lambda < 0. \end{cases}$$

Example 1.4. The sum of independent trapezoidal fuzzy variables $\xi = (a_1, a_2, a_3, a_4)$ and $\eta = (b_1, b_2, b_3, b_4)$ is also a trapezoidal fuzzy variable,

$$\xi + \eta = (a_1 + b_1, a_2 + b_2, a_3 + b_3, a_4 + b_4).$$

The product of a trapezoidal fuzzy variable $\xi = (a_1, a_2, a_3, a_4)$ and a scalar number λ is also a trapezoidal fuzzy variable, and

$$\lambda \cdot \xi = \begin{cases} (\lambda a_1, \lambda a_2, \lambda a_3, \lambda a_4), & \text{if } \lambda \geq 0 \\ (\lambda a_4, \lambda a_3, \lambda a_2, \lambda a_1), & \text{if } \lambda < 0. \end{cases}$$

1.2.3 Expected Value

Definition 1.18 (Liu and Liu [19]). Let ξ be a fuzzy variable. Then the expected value of ξ is defined by

$$E[\xi] = \int_0^{+\infty} \text{Cr}\{\xi \geq r\} dr - \int_{-\infty}^0 \text{Cr}\{\xi \leq r\} dr$$

provided that at least one of the two integrals is finite.

Example 1.5. Let ξ be an equipossible fuzzy variable (a,b). The expected value is $E[\xi] = (a+b)/2$.

Example 1.6. The triangular fuzzy variable $\xi = (a,b,c)$ has an expected value

$$E[\xi] = \frac{1}{4}(a + 2b + c).$$

Example 1.7. The expected value of a trapezoidal fuzzy variable $\xi = (a,b,c,d)$ is

$$E[\xi] = \frac{1}{4}(a + b + c + d).$$

Definition 1.19 (Liu [19]). Let ξ be a fuzzy variable with finite expected value e. Then the variance of ξ is

$$V[\xi] = E\left[(\xi - e)^2\right].$$

Theorem 1.22. *If ξ is a fuzzy variable whose variance exists, a and b are real numbers, then $V[a\xi + b] = a^2 V[\xi]$.*

Proof. It follows from the definition of variance that

$$V[a\xi + b] = E\left[(a\xi + b - aE[\xi] - b)^2\right] = a^2 E\left[(\xi - E[\xi])^2\right] = a^2 V[\xi].$$

1.3 Uncertainty Theory

In order to have a better mathematical tool to deal with empirical data, uncertainty theory was founded by Liu [7] in 2007 and refined in 2010 [11]. As extensions of uncertainty theory, uncertain process and uncertain differential equations (Liu [8]) and uncertain calculus (Liu [9]) were proposed. Uncertain programming was first proposed by Liu [10] in 2009, which wants to deal with the optimal problems involving uncertain variable. This work was followed by an uncertain multiobjective programming and an uncertain goal programming (Liu and Chen [16]) and an uncertain multilevel programming (Liu and Yao [20]). Since that, uncertainty theory was used to solve a variety of real optimal problems, including finance (Chen and Liu [1], Peng [22], Liu [13]), reliability analysis (Liu [12], Zeng et al. [25]), and graph (Gao [2], Gao and Gao [3]). This section will introduce the concepts of uncertain measure, uncertain variable, uncertainty distribution, independence, operational law, expected value, and variance.

1.3.1 Uncertain Measure

Let Γ be a nonempty set and $\mathcal{L}$ a σ-algebra over Γ. Each element $\Lambda \in \mathcal{L}$ is assigned a number $\mathcal{M}\{\Lambda\} \in [0, 1]$. In order to ensure that the number $\mathcal{M}\{\Lambda\}$ has certain mathematical properties, Liu [7, 11] presented the three axioms:

(i) $\mathcal{M}\{\Gamma\} = 1$ for the universal set Γ.
(ii) $\mathcal{M}\{\Lambda\} + \mathcal{M}\{\Lambda^c\} = 1$ for any event Λ.
(iii) For every countable sequence of events $\Lambda_1, \Lambda_2, \cdots$, we have

$$\mathcal{M}\left\{\bigcup_{i=1}^{\infty} \Lambda_i\right\} \leq \sum_{i=1}^{\infty} \mathcal{M}\{\Lambda_i\}$$

The triplet $(\Gamma, \mathcal{L}, \mathcal{M})$ is called an uncertainty space. In order to obtain an uncertain measure of compound event, a product uncertain measure was defined by Liu [9], thus producing the fourth axiom of uncertainty theory:

(iv) Let $(\Gamma_k, \mathcal{L}_k, \mathcal{M}_k)$ be uncertainty spaces for $k = 1, 2, \cdots, \infty$. Then the product uncertain measure $\mathcal{M}$ is an uncertain measure satisfying

$$\mathcal{M}\left\{\prod_{k=1}^{\infty} \Lambda_k\right\} = \bigwedge_{k=1}^{\infty} \mathcal{M}_k\{\Lambda_k\}.$$

Definition 1.20 (Liu [7]). The set function $\mathcal{M}$ is called an uncertain measure if it satisfies the normality, duality, and subadditivity axioms.

The uncertain measure has the following properties:

(i) $\mathcal{M}\{\emptyset\} = 0$.
(ii) $0 \leq \mathcal{M}\{\Lambda\} \leq 1$ for any event Λ.
(iii) $\mathcal{M}\{\Lambda_1\} \leq \mathcal{M}\{\Lambda_2\}$ for any events $\Lambda_1 \subset \Lambda_2$.

1.3.2 Uncertain Variable

Definition 1.21 (Liu [7]). An uncertain variable is a measurable function ξ from an uncertainty space $(\Gamma, \mathcal{L}, \mathcal{M})$ to the set of real numbers, i.e., for any Borel set B of real numbers, the set

$$\{\xi \in B\} = \{\gamma \in \Gamma \mid \xi(\gamma) \in B\} \tag{1.44}$$

is an event.

1.3 Uncertainty Theory

Definition 1.22 (Liu [7]). The uncertainty distribution Φ of an uncertain variable ξ is defined by

$$\Phi(x) = \mathcal{M}\{\xi \leq x\} \tag{1.45}$$

for any real number x.

Theorem 1.23 (Peng Iwamura Theorem [21]: Sufficient and Necessary Condition). *A function $\Phi : \mathfrak{R} \to [0, 1]$ is an uncertainty distribution if and only if it is a monotone increasing function except $\Phi(x) \equiv 0$ and $\Phi(x) \equiv 1$.*

Proof. It is obvious that an uncertainty distribution Φ is a monotone increasing function. In addition, both $\Phi(x) \not\equiv 0$ and $\Phi(x) \not\equiv 1$ follow from the asymptotic theorem immediately. Conversely, suppose that Φ is a monotone increasing function but $\Phi(x) \not\equiv 0$ and $\Phi(x) \not\equiv 1$. We will prove that there is an uncertain variable whose uncertainty distribution is just Φ. Let $\mathcal{C}$ be a collection of all intervals of the form $(-\infty, a], (b, \infty), \emptyset$ and $\mathfrak{R}$. We define a set function on $\mathfrak{R}$ as follows:

$$\mathcal{M}\{(-\infty, a]\} = \Phi(a),$$
$$\mathcal{M}\{(b, +\infty)\} = 1 - \Phi(b),$$
$$\mathcal{M}\{\emptyset\} = 0, \quad \mathcal{M}\{\mathfrak{R}\} = 1.$$

For an arbitrary Borel set B of real numbers, there exists a sequence $\{A_i\}$ in $\mathcal{C}$ such that

$$B \subset \bigcup_{i=1}^{\infty} A_i.$$

Note that such a sequence is not unique. Thus the set function $\mathcal{M}\{B\}$ is defined by

$$\mathcal{M}\{B\} = \begin{cases} \displaystyle\inf_{B \subset \bigcup_{i=1}^{\infty} A_i} \sum_{i=1}^{\infty} \mathcal{M}\{A_i\}, & \text{if } \displaystyle\inf_{B \subset \bigcup_{i=1}^{\infty} A_i} \sum_{i=1}^{\infty} \mathcal{M}\{A_i\} < 0.5 \\ 1 - \displaystyle\inf_{B^c \subset \bigcup_{i=1}^{\infty} A_i} \sum_{i=1}^{\infty} \mathcal{M}\{A_i\}, & \text{if } \displaystyle\inf_{B^c \subset \bigcup_{i=1}^{\infty} A_i} \sum_{i=1}^{\infty} \mathcal{M}\{A_i\} < 0.5 \\ 0.5, & \text{otherwise.} \end{cases}$$

We may prove that the set function $\mathcal{M}$ is indeed an uncertain measure on $\mathfrak{R}$ and the uncertain variable defined by the identity function $\xi(\gamma) = \gamma$ from the uncertainty space $(\mathfrak{R}, \mathcal{L}, \mathcal{M})$ to $\mathfrak{R}$ has the uncertainty distribution Φ.

Example 1.8. An uncertain variable ξ is called linear if it has a linear uncertainty distribution

$$\Phi(x) = \begin{cases} 0, & \text{if } x \leq a \\ (x-a)/(b-a), & \text{if } a \leq x \leq b \\ 1, & \text{if } x \geq b \end{cases} \quad (1.46)$$

denoted by $\mathcal{L}(a,b)$ where a and b are real numbers with $a < b$.

Example 1.9. An uncertain variable ξ is called zigzag if it has a zigzag uncertainty distribution

$$\Phi(x) = \begin{cases} 0, & \text{if } x \leq a \\ (x-a)/2(b-a), & \text{if } a \leq x \leq b \\ (x+c-2b)/2(c-b), & \text{if } b \leq x \leq c \\ 1, & \text{if } x \geq c \end{cases} \quad (1.47)$$

denoted by $\mathcal{Z}(a,b,c)$ where a,b,c are real numbers with $a < b < c$.

Example 1.10. An uncertain variable ξ is called normal if it has a normal uncertainty distribution

$$\Phi(x) = \left(1 + \exp\left(\frac{\pi(e-x)}{\sqrt{3}\sigma}\right)\right)^{-1}, \quad x \in \Re \quad (1.48)$$

denoted by $\mathcal{N}(e,\sigma)$ where e and σ are real numbers with $\sigma > 0$.

Example 1.11. An uncertain variable ξ is called lognormal if $\ln \xi$ is a normal uncertain variable $\mathcal{N}(e,\sigma)$. In other words, a lognormal uncertain variable has an uncertainty distribution

$$\Phi(x) = \left(1 + \exp\left(\frac{\pi(e-\ln x)}{\sqrt{3}\sigma}\right)\right)^{-1}, \quad x \geq 0 \quad (1.49)$$

denoted by $\mathcal{LOGN}(e,\sigma)$, where e and σ are real numbers with $\sigma > 0$.

Example 1.12. An uncertain variable ξ is said to have an empirical uncertainty distribution if

$$\Phi(x) = \begin{cases} 0, & \text{if } x < x_1 \\ \alpha_i + \dfrac{(\alpha_{i+1}-\alpha_i)(x-x_i)}{x_{i+1}-x_i}, & \text{if } x_i \leq x \leq x_{i+1},\ 1 \leq i < n \\ 1, & \text{if } x > x_n \end{cases} \quad (1.50)$$

denoted by $\mathcal{E}(x_1,\alpha_1,x_2,\alpha_2,\cdots,x_n,\alpha_n)$, where $x_1 < x_2 < \cdots < x_n$ and $0 \leq \alpha_1 \leq \alpha_2 \leq \cdots \leq \alpha_n \leq 1$.

1.3 Uncertainty Theory

Definition 1.23. An uncertainty distribution Φ is said to be regular if its inverse function $\Phi^{-1}(\alpha)$ exists and is unique for each $\alpha \in (0, 1)$.

For example, linear uncertainty distribution, zigzag uncertainty distribution, normal uncertainty distribution, and lognormal uncertainty distribution are all regular.

Definition 1.24. Let ξ be an uncertain variable with regular uncertainty distribution Φ. Then the inverse function Φ^{-1} is called the inverse uncertainty distribution of ξ.

Example 1.13. The inverse uncertainty distribution of zigzag uncertain variable $\mathcal{Z}(a, b, c)$ is

$$\Phi^{-1}(\alpha) = \begin{cases} (1 - 2\alpha)a + 2\alpha b, & \text{if } \alpha < 0.5 \\ (2 - 2\alpha)b + (2\alpha - 1)c, & \text{if } \alpha \geq 0.5. \end{cases} \quad (1.51)$$

Example 1.14. The inverse uncertainty distribution of normal uncertain variable $\mathcal{N}(e, \sigma)$ is

$$\Phi^{-1}(\alpha) = e + \frac{\sigma\sqrt{3}}{\pi} \ln \frac{\alpha}{1 - \alpha}. \quad (1.52)$$

Example 1.15. The inverse uncertainty distribution of lognormal uncertain variable $\mathcal{LOGN}(e, \sigma)$ is

$$\Phi^{-1}(\alpha) = \exp(e) \left(\frac{\alpha}{1 - \alpha}\right)^{\sqrt{3}\sigma/\pi}. \quad (1.53)$$

1.3.3 Operational Law

Definition 1.25 (Liu [9]). The uncertain variables $\xi_1, \xi_2, \cdots, \xi_n$ are said to be independent if

$$\mathcal{M}\left\{\bigcap_{i=1}^{n}(\xi_i \in B_i)\right\} = \bigwedge_{i=1}^{n} \mathcal{M}\{\xi_i \in B_i\} \quad (1.54)$$

for any Borel sets $B_1, B_2, \cdots, B_n$ of real numbers.

Theorem 1.24 (Liu [9]). *The uncertain variables $\xi_1, \xi_2, \cdots, \xi_n$ are independent if and only if*

$$\mathcal{M}\left\{\bigcup_{i=1}^{n}(\xi_i \in B_i)\right\} = \bigvee_{i=1}^{n} \mathcal{M}\{\xi_i \in B_i\} \quad (1.55)$$

for any Borel sets $B_1, B_2, \cdots, B_n$ of real numbers.

Proof. It follows from the duality of uncertain measure that $\xi_1, \xi_2, \cdots, \xi_n$ are independent if and only if

$$\mathcal{M}\left\{\bigcup_{i=1}^{n}(\xi_i \in B_i)\right\} = 1 - \mathcal{M}\left\{\bigcap_{i=1}^{n}(\xi_i \in B_i^c)\right\}$$

$$= 1 - \bigwedge_{i=1}^{n}\mathcal{M}\{\xi_i \in B_i^c\} = \bigvee_{i=1}^{n}\mathcal{M}\{\xi_i \in B_i\}.$$

Thus the proof is complete.

Theorem 1.25 (Liu [9]). *Let $\xi_1, \xi_2, \cdots, \xi_n$ be independent uncertain variables, and $f_1, f_2, \cdots, f_n$ measurable functions. Then $f_1(\xi_1), f_2(\xi_2), \cdots, f_n(\xi_n)$ are independent uncertain variables.*

Proof. For any Borel sets $B_1, B_2, \cdots, B_n$ of real numbers, it follows from the definition of independence that

$$\mathcal{M}\left\{\bigcap_{i=1}^{n}(f_i(\xi_i) \in B_i)\right\} = \mathcal{M}\left\{\bigcap_{i=1}^{n}(\xi_i \in f_i^{-1}(B_i))\right\}$$

$$= \bigwedge_{i=1}^{n}\mathcal{M}\{\xi_i \in f_i^{-1}(B_i)\} = \bigwedge_{i=1}^{n}\mathcal{M}\{f_i(\xi_i) \in B_i\}.$$

Thus $f_1(\xi_1), f_2(\xi_2), \cdots, f_n(\xi_n)$ are independent uncertain variables.

A real-valued function $f(x_1, x_2, \cdots, x_n)$ is said to be strictly monotone if it is strictly increasing with respect to $x_1, x_2, \cdots, x_m$ and strictly decreasing with respect to $x_{m+1}, x_{m+2}, \cdots, x_n$, that is,

$$f(x_1, \cdots, x_m, x_{m+1}, \cdots, x_n) \leq f(y_1, \cdots, y_m, y_{m+1}, \cdots, y_n) \quad (1.56)$$

whenever $x_i \leq y_i$ for $i = 1, 2, \cdots, m$ and $x_i \geq y_i$ for $i = m+1, m+2, \cdots, n$, and

$$f(x_1, \cdots, x_m, x_{m+1}, \cdots, x_n) < f(y_1, \cdots, y_m, y_{m+1}, \cdots, y_n) \quad (1.57)$$

whenever $x_i < y_i$ for $i = 1, 2, \cdots, m$ and $x_i > y_i$ for $i = m+1, m+2, \cdots, n$. The following are strictly monotone functions:

$$f(x_1, x_2) = x_1 - x_2,$$
$$f(x_1, x_2) = x_1/x_2, \quad x_1, x_2 > 0.$$

Theorem 1.26 (Liu [11]). *Let $\xi_1, \xi_2, \cdots, \xi_n$ be independent uncertain variables with regular uncertainty distributions $\Phi_1, \Phi_2, \cdots, \Phi_n$, respectively. If the function*

$f(x_1, x_2, \cdots, x_n)$ *is strictly increasing with respect to* $x_1, x_2, \cdots, x_m$ *and strictly decreasing with respect to* $x_{m+1}, x_{m+2}, \cdots, x_n$, *then*

$$\xi = f(\xi_1, \xi_2, \cdots, \xi_n)$$

is an uncertain variable with inverse uncertainty distribution

$$\Psi^{-1}(\alpha) = f(\Phi_1^{-1}(\alpha), \cdots, \Phi_m^{-1}(\alpha), \Phi_{m+1}^{-1}(1-\alpha), \cdots, \Phi_n^{-1}(1-\alpha)). \quad (1.58)$$

Proof. We only prove the case of $m = 1$ and $n = 2$. At first, we always have

$$\mathcal{M}\{\xi \leq \Psi^{-1}(\alpha)\} = \mathcal{M}\{f(\xi_1, \xi_2) \leq f(\Phi_1^{-1}(\alpha), \Phi_2^{-1}(1-\alpha))\}.$$

Since the function $f(x_1, x_2)$ is strictly increasing with respect to x_1 and strictly decreasing with x_2, we obtain

$$\{\xi \leq \Psi^{-1}(\alpha)\} \supset \{\xi_1 \leq \Phi_1^{-1}(\alpha)\} \cap \{\xi_2 \geq \Phi_2^{-1}(1-\alpha)\}.$$

By using the independence of ξ_1 and ξ_2, we get

$$\mathcal{M}\{\xi \leq \Psi^{-1}(\alpha)\} \geq \mathcal{M}\{\xi_1 \leq \Phi_1^{-1}(\alpha)\} \wedge \mathcal{M}\{\xi_2 \geq \Phi_2^{-1}(1-\alpha)\} = \alpha \wedge \alpha = \alpha.$$

On the other hand, since the function $f(x_1, x_2)$ is strictly increasing with respect to x_1 and strictly decreasing with x_2, we obtain

$$\{\xi \leq \Psi^{-1}(\alpha)\} \subset \{\xi_1 \leq \Phi_1^{-1}(\alpha)\} \cup \{\xi_2 \geq \Phi_2^{-1}(1-\alpha)\}.$$

By using the independence of ξ_1 and ξ_2, we get

$$\mathcal{M}\{\xi \leq \Psi^{-1}(\alpha)\} \leq \mathcal{M}\{\xi_1 \leq \Phi_1^{-1}(\alpha)\} \vee \mathcal{M}\{\xi_2 \geq \Phi_2^{-1}(1-\alpha)\} = \alpha \vee \alpha = \alpha.$$

It follows that $\mathcal{M}\{\xi \leq \Psi^{-1}(\alpha)\} = \alpha$. In other words, Ψ is just the uncertainty distribution of ξ. The theorem is proved.

Example 1.16. Let ξ_1 and ξ_2 be independent uncertain variables with regular uncertainty distributions Φ_1 and Φ_2, respectively. Since the function

$$f(x_1, x_2) = x_1 - x_2$$

is strictly increasing with respect to x_1 and strictly decreasing with respect to x_2, the inverse uncertainty distribution of the difference $\xi_1 - \xi_2$ is

$$\Psi^{-1}(\alpha) = \Phi_1^{-1}(\alpha) - \Phi_2^{-1}(1-\alpha). \quad (1.59)$$

Example 1.17. Assume ξ_1, ξ_2, ξ_3 are independent and positive uncertain variables with regular uncertainty distributions Φ_1, Φ_2, Φ_3, respectively. Since the function

$$f(x_1, x_2, x_3) = x_1/(x_2 + x_3)$$

is strictly increasing with respect to x_1 and strictly decreasing with respect to x_2 and x_3, the inverse uncertainty distribution of $\xi_1/(\xi_2 + \xi_3)$ is

$$\Psi^{-1}(\alpha) = \Phi_1^{-1}(\alpha)/(\Phi_2^{-1}(1-\alpha) + \Phi_3^{-1}(1-\alpha)). \tag{1.60}$$

Theorem 1.27 (Liu [11]). *Let $\xi_1, \xi_2, \cdots, \xi_n$ be independent uncertain variables with continuous uncertainty distributions $\Phi_1, \Phi_2, \cdots, \Phi_n$, respectively. If the function $f(x_1, x_2, \cdots, x_n)$ is strictly increasing with respect to $x_1, x_2, \cdots, x_m$ and strictly decreasing with respect to $x_{m+1}, x_{m+2}, \cdots, x_n$, then*

$$\xi = f(\xi_1, \xi_2, \cdots, \xi_n) \tag{1.61}$$

is an uncertain variable with uncertainty distribution

$$\Psi(x) = \sup_{f(x_1, x_2, \cdots, x_n) = x} \left(\min_{1 \leq i \leq m} \Phi_i(x_i) \wedge \min_{m+1 \leq i \leq n} (1 - \Phi_i(x_i)) \right). \tag{1.62}$$

Proof. For simplicity, we only prove the case of $m = 1$ and $n = 2$. Since $f(x_1, x_2)$ is strictly increasing with respect to x_1 and strictly decreasing with respect to x_2, it follows from the definition of uncertainty distribution that

$$\Psi(x) = \mathcal{M}\{f(\xi_1, \xi_2) \leq x\} = \mathcal{M}\left\{ \bigcup_{f(x_1, x_2) = x} (\xi_1 \leq x_1) \cap (\xi_2 \geq x_2) \right\}.$$

Note that for each given x, the event

$$\bigcup_{f(x_1, x_2) = x} (\xi_1 \leq x_1) \cap (\xi_2 \geq x_2)$$

is just a polyrectangle. It follows from the polyrectangular theorem that

$$\Psi(x) = \sup_{f(x_1, x_2) = x} \mathcal{M}\{(\xi_1 \leq x_1) \cap (\xi_2 \geq x_2)\}$$

$$= \sup_{f(x_1, x_2) = x} \mathcal{M}\{\xi_1 \leq x_1\} \wedge \mathcal{M}\{\xi_2 \geq x_2\}$$

$$= \sup_{f(x_1, x_2) = x} \Phi_1(x_1) \wedge (1 - \Phi_2(x_2)).$$

The theorem is proved.

1.3 Uncertainty Theory

Example 1.18. Let ξ_1 and ξ_2 be independent uncertain variables with continuous uncertainty distributions Φ_1 and Φ_2, respectively. Then $\xi_1 - \xi_2$ is an uncertain variable with uncertainty distribution

$$\Psi(x) = \sup_{y \in \Re} \Phi_1(x+y) \wedge (1 - \Phi_2(y)). \tag{1.63}$$

Example 1.19. Let ξ_1 and ξ_2 be independent and positive uncertain variables with continuous uncertainty distributions Φ_1 and Φ_2, respectively. Then ξ_1/ξ_2 is an uncertain variable with uncertainty distribution

$$\Psi(x) = \sup_{y > 0} \Phi_1(xy) \wedge (1 - \Phi_2(y)). \tag{1.64}$$

1.3.4 Expected Value

Definition 1.26 (Liu [9]). Let ξ be an uncertain variable. Then the expected value of ξ is defined by

$$E[\xi] = \int_0^{+\infty} \mathcal{M}\{\xi \geq r\} dr - \int_{-\infty}^0 \mathcal{M}\{\xi \leq r\} dr \tag{1.65}$$

provided that at least one of the two integrals is finite.

Theorem 1.28. *Let ξ be an uncertain variable with uncertainty distribution Φ. If the expected value exists, then*

$$E[\xi] = \int_0^{+\infty} (1 - \Phi(x)) dx - \int_{-\infty}^0 \Phi(x) dx. \tag{1.66}$$

Proof. It follows from the definitions of expected value operator and uncertainty distribution that

$$E[\xi] = \int_0^{+\infty} \mathcal{M}\{\xi \geq r\} dr - \int_{-\infty}^0 \mathcal{M}\{\xi \leq r\} dr$$

$$= \int_0^{+\infty} (1 - \Phi(x)) dx - \int_{-\infty}^0 \Phi(x) dx.$$

The theorem is proved.

Theorem 1.29. *Let ξ be an uncertain variable with regular uncertainty distribution Φ. If the expected value exists, then*

$$E[\xi] = \int_0^1 \Phi^{-1}(\alpha) d\alpha. \tag{1.67}$$

Proof. It follows from the definitions of expected value operator and uncertainty distribution that

$$E[\xi] = \int_0^{+\infty} (1 - \Phi(x))dx - \int_{-\infty}^0 \Phi(x)dx$$

$$= \int_{\Phi(0)}^1 \Phi^{-1}(\alpha)d\alpha + \int_0^{\Phi(0)} \Phi^{-1}(\alpha)d\alpha = \int_0^1 \Phi^{-1}(\alpha)d\alpha.$$

The theorem is proved.

Example 1.20. Let $\xi \sim \mathcal{L}(a,b)$ be a linear uncertain variable. Then its inverse uncertainty distribution is $\Phi^{-1}(\alpha) = (1-\alpha)a + \alpha b$, and its expected value is

$$E[\xi] = \int_0^1 ((1-\alpha)a + \alpha b)d\alpha = \frac{a+b}{2}.$$

Example 1.21. The zigzag uncertain variable $\xi \sim \mathcal{Z}(a,b,c)$ has an expected value

$$E[\xi] = \frac{a + 2b + c}{4}. \tag{1.68}$$

Example 1.22. The normal uncertain variable $\xi \sim \mathcal{N}(e, \sigma)$ has an expected value e, i.e.,

$$E[\xi] = e. \tag{1.69}$$

Example 1.23. If $\sigma < \pi/\sqrt{3}$, then the lognormal uncertain variable $\xi \sim \mathcal{LOGN}(e, \sigma)$ has an expected value

$$E[\xi] = \sqrt{3}\sigma \exp(e) \csc\left(\sqrt{3}\sigma\right). \tag{1.70}$$

Otherwise, $E[\xi] = +\infty$.

Example 1.24. Let ξ have an empirical uncertainty distribution, i.e., $\xi \sim \mathcal{E}(x_1, \alpha_1, x_2, \alpha_2, \cdots, x_n, \alpha_n)$. Then

$$E[\xi] = \frac{\alpha_1 + \alpha_2}{2}x_1 + \sum_{i=2}^{n-1} \frac{\alpha_{i+1} - \alpha_{i-1}}{2}x_i + \left(1 - \frac{\alpha_{n-1} + \alpha_n}{2}\right)x_n \tag{1.71}$$

where $x_1 < x_2 < \cdots < x_m$ and $0 \leq \alpha_1 \leq \alpha_2 \leq \cdots \leq \alpha_n \leq 1$.

1.3 Uncertainty Theory

Example 1.25. Let $\xi \sim \mathcal{D}(x_1, \alpha_1, x_2, \alpha_2, \cdots, x_m, \alpha_m)$ be a discrete uncertain variable. Then

$$E[\xi] = \sum_{i=1}^{m}(\alpha_i - \alpha_{i-1})x_i \qquad (1.72)$$

where $x_1 < x_2 < \cdots < x_m$ and $0 = \alpha_0 \leq \alpha_1 \leq \alpha_2 \leq \cdots \leq \alpha_m = 1$.

Theorem 1.30 (Liu and Ha [18]). *Assume $\xi_1, \xi_2, \cdots, \xi_n$ are independent uncertain variables with regular uncertainty distributions $\Phi_1, \Phi_2, \cdots, \Phi_n$, respectively. If $f(x_1, x_2, \cdots, x_n)$ is strictly increasing with respect to $x_1, x_2, \cdots, x_m$ and strictly decreasing with respect to $x_{m+1}, x_{m+2}, \cdots, x_n$, then the uncertain variable $\xi = f(\xi_1, \xi_2, \cdots, \xi_n)$ has an expected value*

$$E[\xi] = \int_0^1 f\left(\Phi_1^{-1}(\alpha), \cdots, \Phi_m^{-1}(\alpha), \Phi_{m+1}^{-1}(1-\alpha), \cdots, \Phi_n^{-1}(1-\alpha)\right) d\alpha \qquad (1.73)$$

provided that $E[\xi]$ exists.

Proof. Since the function $f(x_1, x_2, \cdots, x_n)$ is strictly increasing with respect to $x_1, x_2, \cdots, x_m$ and strictly decreasing with respect to $x_{m+1}, x_{m+2}, \cdots, x_n$, it follows from Theorem 1.26 that the inverse uncertainty distribution of ξ is

$$\Psi^{-1}(\alpha) = f\left(\Phi_1^{-1}(\alpha), \cdots, \Phi_m^{-1}(\alpha), \Phi_{m+1}^{-1}(1-\alpha), \cdots, \Phi_n^{-1}(1-\alpha)\right)$$

By using Theorem 1.29, we obtain (1.73). The theorem is proved.

Example 1.26. Let ξ be a nonnegative uncertain variable with regular uncertainty distribution Φ. Then

$$E\left[\sqrt{\xi}\right] = \int_0^1 \sqrt{\Phi^{-1}(\alpha)}\,d\alpha. \qquad (1.74)$$

Example 1.27. Let ξ be an uncertain variable with regular uncertainty distribution Φ. Then

$$E[\exp(\xi)] = \int_0^1 \exp\left(\Phi^{-1}(\alpha)\right)d\alpha. \qquad (1.75)$$

Example 1.28. Let ξ be a positive uncertain variable with regular uncertainty distribution Φ. Then

$$E\left[\frac{1}{\xi}\right] = \int_0^1 \frac{1}{\Phi^{-1}(1-\alpha)}\,d\alpha = \int_0^1 \frac{1}{\Phi^{-1}(\alpha)}\,d\alpha. \qquad (1.76)$$

Theorem 1.31 (Liu [11]). *Let ξ and η be independent uncertain variables with finite expected values. Then for any real numbers a and b, we have*

$$E[a\xi + b\eta] = aE[\xi] + bE[\eta]. \tag{1.77}$$

Proof. Without loss of generality, suppose ξ and η have regular uncertainty distributions Φ and Ψ, respectively. Otherwise, we may give the uncertainty distributions a small perturbation such that they become regular.

Step 1: We first prove $E[a\xi] = aE[\xi]$. If $a = 0$, then the equation holds trivially. If $a > 0$, then the inverse uncertainty distribution of $a\xi$ is

$$\Upsilon^{-1}(\alpha) = a\Phi^{-1}(\alpha).$$

It follows from Theorem 1.29 that

$$E[a\xi] = \int_0^1 a\Phi^{-1}(\alpha)d\alpha = a\int_0^1 \Phi^{-1}(\alpha)d\alpha = aE[\xi].$$

If $a < 0$, then the inverse uncertainty distribution of $a\xi$ is

$$\Upsilon^{-1}(\alpha) = a\Phi^{-1}(1-\alpha).$$

It follows from Theorem 1.29 that

$$E[a\xi] = \int_0^1 a\Phi^{-1}(1-\alpha)d\alpha = a\int_0^1 \Phi^{-1}(\alpha)d\alpha = aE[\xi].$$

Thus we always have $E[a\xi] = aE[\xi]$.

Step 2: We prove $E[\xi + \eta] = E[\xi] + E[\eta]$. The inverse uncertainty distribution of the sum $\xi + \eta$ is

$$\Upsilon^{-1}(\alpha) = \Phi^{-1}(\alpha) + \Psi^{-1}(\alpha).$$

It follows from Theorem 1.29 that

$$E[\xi + \eta] = \int_0^1 \Upsilon^{-1}(\alpha)d\alpha = \int_0^1 \Phi^{-1}(\alpha)d\alpha + \int_0^1 \Psi^{-1}(\alpha)d\alpha = E[\xi] + E[\eta].$$

Step 3: Finally, for any real numbers a and b, it follows from Steps 1 and 2 that

$$E[a\xi + b\eta] = E[a\xi] + E[b\eta] = aE[\xi] + bE[\eta].$$

The theorem is proved.

1.4 Chance Theory

Definition 1.27 (Liu [9]). Let ξ be an uncertain variable with finite expected value e. Then the variance of ξ is

$$V[\xi] = E\left[(\xi - e)^2\right]. \tag{1.78}$$

This definition tells us that the variance is just the expected value of $(\xi - e)^2$. Since $(\xi - e)^2$ is a nonnegative uncertain variable, we also have

$$V[\xi] = \int_0^{+\infty} \mathcal{M}\left\{(\xi - e)^2 \geq r\right\} dr. \tag{1.79}$$

Theorem 1.32. *If ξ is an uncertain variable with finite expected value, a and b are real numbers, then*

$$V[a\xi + b] = a^2 V[\xi]. \tag{1.80}$$

Proof. Let e be the expected value of ξ. Then $a\xi + b$ has an expected value $ae + b$. It follows from the definition of variance that

$$V[a\xi + b] = E\left[(a\xi + b - (ae + b))^2\right] = a^2 E\left[(\xi - e)^2\right] = a^2 V[\xi].$$

The theorem is thus verified.

1.4 Chance Theory

In many cases, uncertainty and randomness simultaneously appear in a system. For example, some quantities have no samples while others have samples enough to determine probability distributions. In order to describe this phenomenon, the concepts of uncertain random variable and chance measure were pioneered by Liu [14] in 2012. Chance theory begins with uncertain random variable and chance measure and is a mathematical methodology composed of uncertainty theory and probability theory.

1.4.1 Uncertain Random Variable

Roughly speaking, an uncertain random variable is a function from a probability space to a collection of uncertain variables. In other words, an uncertain random variable is a random element taking values of uncertain variables.

Definition 1.28 (Liu [14]). An uncertain random variable is a function ξ from a probability space $(\Omega, \mathcal{A}, \Pr)$ to a collection of uncertain variables such that

$$\mathcal{M}\{\xi(\omega) \in B\} \tag{1.81}$$

is a measurable function of ω for any Borel set B of real numbers.

Example 1.29. A random variable is a special uncertain random variable because any real value is a special uncertain variable.

Example 1.30. An uncertain variable is a special uncertain random variable because it is a function from a probability space to the uncertain variable itself.

Example 1.31. Let $\eta_1, \eta_2, \cdots, \eta_m$ be random variables and let $\tau_1, \tau_2, \cdots, \tau_n$ be uncertain variables. If f is a measurable function, then

$$\xi = f(\eta_1, \eta_2, \cdots, \eta_m, \tau_1, \tau_2, \cdots, \tau_n) \tag{1.82}$$

is an uncertain random variable. Especially, if η is a random variable and τ is an uncertain variable, then their sum $\eta + \tau$, difference $\eta - \tau$, product $\eta\tau$, and quotient η/τ are instances of uncertain random variable.

1.4.2 Uncertain Random Measure

Definition 1.29 (Liu [14]). Let ξ be an uncertain random variable and let B be a Borel set of real numbers. Then the chance measure of uncertain random event $\xi \in B$ is defined by

$$\mathrm{Ch}\{\xi \in B\} = \int_0^1 \Pr\{\omega \in \Omega \mid \mathcal{M}\{\xi(\omega) \in B\} \geq r\} \, \mathrm{d}r. \tag{1.83}$$

Remark 1.1. If an uncertain random variable ξ degenerates to a random variable η, then

$$\mathrm{Ch}\{\xi \in B\} = \Pr\{\eta \in B\}. \tag{1.84}$$

If an uncertain random variable ξ degenerates to an uncertain variable τ, then

$$\mathrm{Ch}\{\xi \in B\} = \mathcal{M}\{\tau \in B\}. \tag{1.85}$$

Remark 1.2. Note that $\mathcal{M}\{\xi(\omega) \in B\}$ is a measurable function of ω and then $\mathcal{M}\{\xi(\cdot) \in B\}$ is a random variable. The chance measure $\mathrm{Ch}\{\xi \in B\}$ is in fact the expected value of the random variable $\mathcal{M}\{\xi(\cdot) \in B\}$, i.e.,

$$\mathrm{Ch}\{\xi \in B\} = E\left[\mathcal{M}\{\xi(\cdot) \in B\}\right]. \tag{1.86}$$

1.4 Chance Theory

Remark 1.3. The chance measure $\text{Ch}\{\xi \in B\}$ may also be written as an abstract integral,

$$\text{Ch}\{\xi \in B\} = \int_\Omega \mathcal{M}\{\xi(\omega) \in B\} d\Pr(\omega). \tag{1.87}$$

The chance measure has the following properties:

(i) $\mathcal{M}\{\emptyset\} = 0 \quad \text{Ch}\{\xi \in \Re\} = 1$.
(ii) $\text{Ch}\{\xi \in B\} + \text{Ch}\{\xi \in B^c\} = 1$ for any event Λ.

Definition 1.30 (Liu [14]). Let ξ be an uncertain random variable. Then its chance distribution is defined by

$$\Phi(x) = \text{Ch}\{\xi \leq x\} \tag{1.88}$$

for any $x \in \Re$.

Theorem 1.33 (Liu [14], Sufficient and Necessary Condition for Chance Distribution). *A function* $\Phi : \Re \to [0, 1]$ *is a chance distribution if and only if it is a monotone increasing function except* $\Phi(x) \equiv 0$ *and* $\Phi(x) \equiv 1$.

Proof. Assume Φ is a chance distribution of uncertain random variable ξ. Let x_1 and x_2 be two real numbers with $x_1 < x_2$. It follows from the monotone increasing theorem that

$$\Phi(x_1) = \text{Ch}\{\xi \leq x_1\} \leq \text{Ch}\{\xi \leq x_2\} = \Phi(x_2).$$

Hence, the chance distribution Φ is a monotone increasing function. Furthermore, if $\Phi(x) \equiv 0$, then

$$\int_0^1 \Pr\{\omega \in \Omega \mid \mathcal{M}\{\xi(\omega) \leq x\} \geq r\} dr \equiv 0.$$

Thus for almost all $\omega \in \Omega$, we have

$$\mathcal{M}\{\xi(\omega) \leq x\} \equiv 0, \quad \forall x \in \Re$$

which is in contradiction to the asymptotic theorem, and then $\Phi(x) \neq 0$ is verified. Similarly, if $\Phi(x) \equiv 1$, then

$$\int_0^1 \Pr\{\omega \in \Omega \mid \mathcal{M}\{\xi(\omega) \leq x\} \geq r\} dr \equiv 1.$$

Thus for almost all $\omega \in \Omega$, we have

$$\mathcal{M}\{\xi(\omega) \leq x\} \equiv 1, \quad \forall x \in \Re$$

which is also in contradiction to the asymptotic theorem, and then $\Phi(x) \neq 1$ is proved.

Theorem 1.34 (Liu [14], Measure Inversion Theorem). *Let ξ be an uncertain random variable with continuous chance distribution Φ. Then for any real number x, we have*

$$\text{Ch}\{\xi \leq x\} = \Phi(x), \quad \text{Ch}\{\xi \geq x\} = 1 - \Phi(x). \tag{1.89}$$

Proof. The equation $\text{Ch}\{\xi \leq x\} = \Phi(x)$ follows from the definition of chance distribution immediately. By using the duality axiom of chance measure and continuity of chance distribution, we get $\text{Ch}\{\xi \geq x\} = 1 - \text{Ch}\{\xi < x\} = 1 - \Phi(x)$.

1.4.3 Operational Law

Definition 1.31 (Liu [15]). The uncertain random variables $\xi_1, \xi_2, \cdots, \xi_n$ are said to be independent if

$$\text{Ch}\left\{\bigcap_{i=1}^{n}(\xi_i \in B_i)\right\} = E\left[\bigwedge_{i=1}^{n}\mathcal{M}\{\xi_i(\cdot) \in B_i\}\right] \tag{1.90}$$

for any Borel sets $B_1, B_2, \cdots, B_n$ of real numbers.

Theorem 1.35. *Let $\xi_1, \xi_2, \cdots, \xi_n$ be independent uncertain random variables and $f_1, f_2, \cdots, f_n$ measurable functions. Then $f_1(\xi_1), f_2(\xi_2), \cdots, f_n(\xi_n)$ are independent uncertain random variables.*

Proof. For any Borel sets $B_1, B_2, \cdots, B_n$ of real numbers, it follows from the definition of independence that

$$\text{Ch}\left\{\bigcap_{i=1}^{n}(f_i(\xi_i) \in B_i)\right\} = \text{Ch}\left\{\bigcap_{i=1}^{n}(\xi_i \in f^{-1}(B_i))\right\}$$

$$= E\left[\bigwedge_{i=1}^{n}\mathcal{M}\{\xi_i(\cdot) \in f^{-1}(B_i)\}\right] = E\left[\bigwedge_{i=1}^{n}\mathcal{M}\{f(\xi_i(\cdot)) \in B_i\}\right].$$

Thus $f_1(\xi_1), f_2(\xi_2), \cdots, f_n(\xi_n)$ are independent uncertain random variables.

Theorem 1.36 (Liu [15]). *Let $\eta_1, \eta_2, \cdots, \eta_m$ be independent random variables with probability distributions $\Psi_1, \Psi_2, \cdots, \Psi_m$, and let $\tau_1, \tau_2, \cdots, \tau_n$ be independent uncertain variables with uncertainty distributions $\Upsilon_1, \Upsilon_2, \cdots, \Upsilon_n$, respectively. Then the uncertain random variable*

$$\xi = f(\eta_1, \eta_2, \cdots, \eta_m, \tau_1, \tau_2, \cdots, \tau_n) \tag{1.91}$$

1.4 Chance Theory

has a chance distribution

$$\Phi(x) = \int_{\Re^m} F(x; y_1, y_2, \cdots, y_m) d\Psi_1(y_1) d\Psi_2(y_2) \cdots d\Psi_m(y_m) \quad (1.92)$$

where $F(x; y_1, y_2, \cdots, y_m)$ is determined by its inverse function

$$F^{-1}(\alpha; y_1, y_2, \cdots, y_m) = f(y_1, y_2, \cdots, y_m, \Upsilon_1^{-1}(\alpha), \Upsilon_2^{-1}(\alpha), \cdots, \Upsilon_n^{-1}(\alpha))$$

provided that $f(\eta_1, \eta_2, \cdots, \eta_m, \tau_1, \tau_2, \cdots, \tau_n)$ is a strictly increasing function with respect to $\tau_1, \tau_2, \cdots, \tau_n$.

Proof. For any given numbers $y_1, y_2, \cdots, y_m$, it follows from the operational law of uncertain variables that $f(y_1, y_2, \cdots, y_m, \tau_1, \tau_2, \cdots, \tau_n)$ is an uncertain variable with uncertainty distribution $F(x; y_1, y_2, \cdots, y_m)$ that is determined by its inverse function $F^{-1}(\alpha; y_1, y_2, \cdots, y_m)$. By using the definition of chance measure, the chance distribution of ξ is

$$\Phi(x) = E[F(x; \eta_1, \eta_2, \cdots, \eta_m)]$$

that is just (1.92). The theorem is verified.

Remark 1.4. If $f(\eta_1, \eta_2, \cdots, \eta_m, \tau_1, \tau_2, \cdots, \tau_n)$ is strictly increasing with respect to $\tau_1, \cdots, \tau_k$ and strictly decreasing with respect to $\tau_{k+1}, \cdots, \tau_n$, then $F^{-1}(\alpha; y_1, y_2, \cdots, y_m)$ is equal to

$$f\left(y_1, y_2, \cdots, y_m, \Upsilon_1^{-1}(\alpha), \cdots, \Upsilon_k^{-1}(\alpha), \Upsilon_{k+1}^{-1}(1-\alpha), \cdots, \Upsilon_n^{-1}(1-\alpha)\right).$$

Example 1.32. Let η be a random variable with probability distribution Ψ and let τ be an uncertain variable with uncertainty distribution Υ. Then the sum

$$\xi = \eta + \tau \quad (1.93)$$

is an uncertain random variable whose chance distribution is

$$\Phi(x) = \int_{-\infty}^{+\infty} \Upsilon(x - y) d\Psi(y). \quad (1.94)$$

Example 1.33. Let η be a random variable with probability distribution Ψ and let τ be an uncertain variable with uncertainty distribution Υ. Then the minimum

$$\xi = \eta \wedge \tau \quad (1.95)$$

is an uncertain random variable whose chance distribution is

$$\Phi(x) = \Psi(x) + \Upsilon(x) - \Psi(x)\Upsilon(x). \quad (1.96)$$

1.4.4 Expected Value

Definition 1.32 (Liu [14]). Let ξ be an uncertain random variable. Then its expected value is defined by

$$E[\xi] = \int_0^{+\infty} \mathrm{Ch}\{\xi \geq r\}\mathrm{d}r - \int_{-\infty}^0 \mathrm{Ch}\{\xi \leq r\}\mathrm{d}r \tag{1.97}$$

provided that at least one of the two integrals is finite.

Theorem 1.37 (Liu [14]). *Let ξ be an uncertain random variable with chance distribution Φ. If the expected value of ξ exists, then*

$$E[\xi] = \int_0^{+\infty} (1 - \Phi(x))\mathrm{d}x - \int_{-\infty}^0 \Phi(x)\mathrm{d}x. \tag{1.98}$$

Proof. It follows from the relation between chance distribution and chance measure that

$$\mathrm{Ch}\{\xi > x\} = 1 - \Phi(x), \quad \mathrm{Ch}\{\xi \leq x\} = \Phi(x)$$

for any real number x. Thus Eq. (1.98) follows from the definition of expected value immediately.

Example 1.34. Let η be a random variable and let τ be an uncertain variable. Assume η has a probability distribution Ψ. It follows from Theorem 1.39 that the uncertain random variable $\eta + \tau$ has an expected value

$$E[\eta + \tau] = \int_\Re E[y + \tau]\mathrm{d}\Psi(y) = \int_\Re (y + E[\tau])\mathrm{d}\Psi(y) = E[\eta] + E[\tau].$$

That is,

$$E[\eta + \tau] = E[\eta] + E[\tau]. \tag{1.99}$$

Example 1.35. Let η be a random variable and let τ be an uncertain variable. Assume η has a probability distribution Ψ. It follows from Theorem 1.39 that the uncertain random variable $\eta\tau$ has an expected value

$$E[\eta\tau] = \int_\Re E[y\tau]\mathrm{d}\Psi(y) = \int_\Re yE[\tau]\mathrm{d}\Psi(y) = E[\eta]E[\tau].$$

That is,

$$E[\eta\tau] = E[\eta]E[\tau]. \tag{1.100}$$

1.4 Chance Theory

Theorem 1.38 (Yao). *Let ξ be an uncertain random variable with finite expected value. For each $\omega \in \Omega$, the realization $\xi(\omega)$ is an uncertain variable whose expected value is denoted by $E[\xi(\omega)]$. Then ξ has an expected value*

$$E[\xi] = \int_\Omega E[\xi(\omega)] \mathrm{d}\mathrm{Pr}(\omega). \tag{1.101}$$

Proof. It follows from the definition of chance measure that

$$\mathrm{Ch}\{\xi \geq x\} = \int_\Omega \mathcal{M}\{\xi(\omega) \geq x\} \mathrm{d}\mathrm{Pr}(\omega),$$

$$\mathrm{Ch}\{\xi \leq x\} = \int_\Omega \mathcal{M}\{\xi(\omega) \leq x\} \mathrm{d}\mathrm{Pr}(\omega).$$

Then

$$E[\xi] = \int_0^{+\infty} \int_\Omega \mathcal{M}\{\xi(\omega) \geq x\} \mathrm{d}\mathrm{Pr}(\omega) \mathrm{d}x - \int_{-\infty}^0 \int_\Omega \mathcal{M}\{\xi(\omega) \leq x\} \mathrm{d}\mathrm{Pr}(\omega) \mathrm{d}x.$$

By using Fubini theorem, we obtain

$$E[\xi] = \int_\Omega \int_0^{+\infty} \mathcal{M}\{\xi(\omega) \geq x\} \mathrm{d}x \mathrm{d}\mathrm{Pr}(\omega) - \int_\Omega \int_{-\infty}^0 \mathcal{M}\{\xi(\omega) \leq x\} \mathrm{d}x \mathrm{d}\mathrm{Pr}(\omega)$$

$$= \int_\Omega \left(\int_0^{+\infty} \mathcal{M}\{\xi(\omega) \geq x\} \mathrm{d}x - \int_{-\infty}^0 \mathcal{M}\{\xi(\omega) \leq x\} \mathrm{d}x \right) \mathrm{d}\mathrm{Pr}(\omega)$$

$$= \int_\Omega E[\xi(\omega)] \mathrm{d}\mathrm{Pr}(\omega).$$

The theorem is proved.

Theorem 1.39 (Liu [15]). *Let $\eta_1, \eta_2, \cdots, \eta_m$ be independent random variables with probability distributions $\Psi_1, \Psi_2, \cdots, \Psi_m$, respectively. Then the uncertain random variable*

$$\xi = f(\eta_1, \cdots, \eta_m, \tau_1, \cdots, \tau_n) \tag{1.102}$$

has an expected value

$$E[\xi] = \int_{\Re^m} E[f(y_1, \cdots, y_m, \tau_1, \cdots, \tau_n)] \mathrm{d}\Psi_1(y_1) \cdots \mathrm{d}\Psi_m(y_m) \tag{1.103}$$

where $E[f(y_1, \cdots, y_m, \tau_1, \cdots, \tau_n)]$ is the expected value of the uncertain variable $f(y_1, \cdots, y_m, \tau_1, \cdots, \tau_n)$ for any given real numbers $y_1, \cdots, y_m$.

Proof. Since $(y_1, \cdots, y_m, \tau_1, \cdots, \tau_n)$ is a realization of $(\eta_1, \cdots, \eta_m, \tau_1, \cdots, \tau_n)$, it follows from Theorem 1.38 that

$$E[\xi] = \int_{\Re^m} E[f(y_1, \cdots, y_m, \tau_1, \cdots, \tau_n)] \mathrm{dPr}(y_1) \cdots \mathrm{dPr}(y_m)$$

$$= \int_{\Re^m} E[f(y_1, \cdots, y_m, \tau_1, \cdots, \tau_n)] \mathrm{d}\Psi_1(y_1) \cdots \mathrm{d}\Psi_m(y_m)$$

The theorem is proved.

Theorem 1.40 (Liu [15]). *Let $\eta_1, \eta_2, \cdots, \eta_m$ be independent random variables with probability distributions $\Psi_1, \Psi_2, \cdots, \Psi_m$, respectively. Then the uncertain random variable*

$$\xi = f(\eta_1, \cdots, \eta_m, \tau_1, \cdots, \tau_n) \tag{1.104}$$

has an expected value

$$E[\xi] = \int_{\Re^m} E[f(y_1, \cdots, y_m, \tau_1, \cdots, \tau_n)] \mathrm{d}\Psi_1(y_1) \cdots \mathrm{d}\Psi_m(y_m) \tag{1.105}$$

where $E[f(y_1, \cdots, y_m, \tau_1, \cdots, \tau_n)]$ is the expected value of the uncertain variable $f(y_1, \cdots, y_m, \tau_1, \cdots, \tau_n)$ for any given real numbers $y_1, \cdots, y_m$.

Proof. Since $f(y_1, \cdots, y_m, \tau_1, \cdots, \tau_n)$ is a strictly increasing function or a strictly decreasing function with respect to $\tau_1, \cdots, \tau_n$, we have

$$E[f(y_1, \cdots, y_m, \tau_1, \cdots, \tau_n)] = \int_0^1 f\left(y_1, \cdots, y_m, \Upsilon_1^{-1}(\alpha), \cdots, \Upsilon_n^{-1}(\alpha)\right) \mathrm{d}\alpha.$$

It follows from Theorem 1.39 that the result holds.

Theorem 1.41. *Let ξ be an uncertain random variable whose expected value exists. If a and b are real numbers, then*

$$E[a\xi + b] = aE[\xi] + b. \tag{1.106}$$

Proof. Step 1: We first prove $E[\xi + b] = E[\xi] + b$ for any real number b. If $b \geq 0$, we have

$$E[\xi + b] = \int_0^{+\infty} \mathrm{Ch}\{\xi + b \geq r\} \mathrm{d}r - \int_{-\infty}^0 \mathrm{Ch}\{\xi + b \leq r\} \mathrm{d}r$$

$$= \int_0^{+\infty} \mathrm{Ch}\{\xi \geq r - b\} \mathrm{d}r - \int_{-\infty}^0 \mathrm{Ch}\{\xi \leq r - b\} \mathrm{d}r$$

$$= E[\xi] + \int_0^b (\mathrm{Ch}\{\xi \geq r - b\} + \mathrm{Ch}\{\xi < r - b\}) \mathrm{d}r$$

$$= E[\xi] + b.$$

Similarly, if $b < 0$, then we have

$$E[\xi + b] = E[\xi] - \int_b^0 (\text{Ch}\{\xi \geq r - b\} + \text{Ch}\{\xi < r - b\})\,dr = E[\xi] + b.$$

Step 2: We prove $E[a\xi] = aE[\xi]$ for any real number a. If $a = 0$, then $E[a\xi] = aE[\xi]$ holds trivially. If $a > 0$, then we have

$$E[a\xi] = \int_0^{+\infty} \text{Ch}\{a\xi \geq r\}dr - \int_{-\infty}^0 \text{Ch}\{a\xi \leq r\}dr$$

$$= a\int_0^{+\infty} \text{Ch}\{\xi \geq t\}\,dt - a\int_{-\infty}^0 \text{Ch}\{\xi \leq t\}\,dt = aE[\xi].$$

Similarly, if $a < 0$, then we have

$$E[a\xi] = \int_0^{+\infty} \text{Ch}\{a\xi \geq r\}dr - \int_{-\infty}^0 \text{Ch}\{a\xi \leq r\}dr$$

$$= -a\int_{-\infty}^0 \text{Ch}\{\xi \leq t\}\,dt + a\int_0^{+\infty} \text{Ch}\{\xi \geq t\}\,dt = aE[\xi].$$

Step 3: For any real numbers a and b, it follows from Steps 1 and 2 that the expected value of $a\xi + b$ is

$$E[a\xi + b] = E[a\xi] + E[b] = aE[\xi] + b.$$

The theorem is proved.

Definition 1.33 (Liu [14]). Let ξ be an uncertain random variable with finite expected value e. Then the variance of ξ is

$$V[\xi] = E\left[(\xi - e)^2\right]. \tag{1.107}$$

Since $(\xi - e)^2$ is a nonnegative uncertain random variable, we also have

$$V[\xi] = \int_0^{+\infty} \text{Ch}\left\{(\xi - e)^2 \geq r\right\} dr. \tag{1.108}$$

Theorem 1.42 (Liu [14]). *If ξ is an uncertain random variable with finite expected value, a and b are real numbers, then $V[a\xi + b] = a^2 V[\xi]$.*

Proof. Let e be the expected value of ξ. Then $a\xi + b$ has an expected value $ae + b$. Thus the variance is

$$V[a\xi + b] = E\left[((a\xi + b) - (ae + b))^2\right] = E\left[a^2(\xi - e)^2\right] = a^2 V[\xi].$$

The theorem is thus proved.

References

1. Chen X, Liu B (2010) Existence and uniqueness theorem for uncertain differential equations. Fuzzy Optim Decis Making 9(1):69–81
2. Gao XL (2013) Cycle index of uncertain graph. Inf Int Interdiscip J 16(2A):131–1138
3. Gao XL, Gao Y (2013) Connectedness index of uncertainty graphs. Int J Uncertain Fuzziness Knowl Based Syst 21(1):127–137
4. Li X, Liu B (2006) A sufficient and necessary condition for credibility measures. Int J Uncertain Fuzziness Knowl Based Syst 14(5):527–535
5. Liu B (2002) Theory and practice of uncertain programming. Physica-Verlag, Heidelberg
6. Liu B (2004) Uncertainty theory. Springer, Berlin
7. Liu B (2007) Uncertainty theory, 2nd edn. Springer, Berlin
8. Liu B (2008) Fuzzy process, hybrid process and uncertain process. J Uncertain Syst 2(1):3–16
9. Liu B (2009) Some research problems in uncertainty theory. J Uncertain Syst 3(1):3–10
10. Liu B (2009) Theory and practice of uncertain programming, 2nd edn. Springer, Berlin
11. Liu B (2010) Uncertainty theory: a branch of mathematics for modeling human uncertainty. Springer, Berlin
12. Liu B (2010) Uncertain risk analysis and uncertain reliability analysis. J Uncertain Syst 4(3):163–170
13. Liu B (2013) Extreme value theorems of uncertain process with application to insurance risk model. Soft Comput 17(4):549–556
14. Liu YH (2013) Uncertain random variables: a mixture of uncertainty and randomness. Soft Comput 17(4):625–634
15. Liu YH (2013) Uncertain random programming with applications. Fuzzy Optim Decis Mak 12:153–169
16. Liu B, Chen XW (2013) Uncertain multiobjective programming and uncertain goal programming. Technical report
17. Liu YH, Gao J (2007) The independence of fuzzy variables with applications to fuzzy random optimization. Int J Uncertain Fuzziness Knowl Based Syst 15(Supp 2):1–20
18. Liu YH, Ha MH (2010) Expected value of function of uncertain variables. J Uncertain Syst 13:181–186
19. Liu B, Liu YK (2002) Expected value of fuzzy variable and fuzzy expected value models. IEEE Trans Fuzzy Syst 10(4):445–450
20. Liu B, Yao K, Uncertain multilevel programming: algorithm and application. http://orsc.edu.cn/online/120114.pdf.
21. Peng ZX, Iwamura K (2010) A sufficient and necessary condition of uncertainty distribution. J Interdiscip Math 13(3):277–285
22. Peng J, Yao K (2010) A new option pricing model for stocks in uncertainty markets. Int J Oper Res 7(4):213–224
23. Zadeh LA (1965) Fuzzy sets. Inf Control 8:338–353
24. Zadeh LA (1978) Fuzzy sets as a basis for a theory of possibility. Fuzzy Sets Syst 1:3–28
25. Zeng ZG, Wen ML, Kang R (2013) Belief reliability: a new metrics for products' reliability. Fuzzy Optim Decis Mak 12(1):15–27

Chapter 2
Introduction to DEA

Data envelopment analysis (DEA), a "data-oriented" approach to evaluate the performance of a set of peer entities, has been widely used since it was first invented by Charnes. This is followed by a series of theoretical extensions. See Banker et al. [1], Charnes et al. [3], Petersen [12], Tone [14], and Cooper [6].

Our focus in this chapter is on basic DEA models for measuring the efficiency of a DMU relative to similar DMUs in order to estimate a "best practice" frontier. The initial DEA model, originally presented in Charnes et al. [2], was built on the earlier work of Farrell [10]. After that, more than 4,000 relevant articles have been published. Such rapid growth and widespread acceptance of the methodology of DEA is testimony to its strength and applicability. Researchers in a number of fields have quickly recognized that DEA is an excellent methodology for modeling operational processes, and its empirical orientation and minimization of a priori assumptions have made possible use in a number of studies involving efficient frontier estimation in the nonprofit sector, the regulated sector, and the private sector.

At present, DEA actually encompasses a variety of alternate approaches to performance evaluation. Extensions to the original CCR work have facilitated a deeper analysis of both the "multiplier side" from the dual model and the "envelopment side" from the primal model of the mathematical duality structure. Properties such as isotonicity; nonconcavity; economies of scale; piecewise linearity; discretionary, categorical variables; and ordinal relationships can also be treated through DEA.

In recent years a great variety of applications of DEA have been proposed. These DEA applications have used DMUs in various forms to evaluate the performance of such entities as hospitals, US Air Force wings, universities, cities, courts, and business firms, as well as the performance of countries, regions, etc.

This chapter will present a literature review on DEA, including the fundamental concept of DEA, frequently used DEA models, and the DMU efficiency definitions.

2.1 Symbols and Notations

In DEA, the organization under study is called a DMU (decision-making unit). The definition of DMU is rather loose to allow flexibility in its use over a wide range of possible applications. Generically a DMU is regarded as the entity responsible for converting inputs into outputs and whose performances are to be evaluated. In managerial applications, DMUs may include banks, department stores, and supermarkets and extend to car makers, hospitals, schools, public libraries, and so on. In engineering, DMUs may take such forms as airplanes or their components such as jet engines. For the purpose of securing relative comparisons, a group of DMUs is used to evaluate each other with each DMU having a certain degree of managerial freedom in decision making.

Suppose there are n DMUs and the symbols and notations are listed as follows:

DMU_i: the ith DMU, $i = 1, 2, \cdots, n$
DMU_0: the target DMU
$x_i = (x_{i1}, x_{i2}, \cdots, x_{ip})$: the inputs vector of DMU_i, $i = 1, 2, \cdots, n$
$x_0 = (x_{01}, x_{02}, \cdots, x_{0p})$: the inputs vector of the target DMU_0
$y_i = (y_{i1}, y_{i2}, \cdots, y_{iq})$: the outputs vector of DMU_i, $i = 1, 2, \cdots, n$
$y_0 = (y_{01}, y_{02}, \cdots, y_{0q})$: the outputs vector of the target DMU_0
$u \in R^{p \times 1}$: the vector of input weights
$v \in R^{q \times 1}$: the vector of output weights

2.2 CCR Model

This section deals with one of the most basic DEA models named CCR model, which was initially proposed by Charnes et al. [2] in 1978:

$$\begin{cases} \max_{u,v} \theta = \dfrac{v^T y_0}{u^T x_0} \\ \text{subject to:} \\ \quad v^T y_j \leq u^T x_j, \ j = 1, 2, \cdots, n \\ \quad u \geq 0 \\ \quad v \geq 0. \end{cases} \quad (2.1)$$

The constraints mean that the ratio of "virtual output" vs. "virtual input" should not exceed 1 for every DMU. The objective is to obtain the ratio of the weighted output to the weighted input weights. By virtue of the constraints, the optimal objective value θ^* is at most 1. Mathematically, the nonnegativity constraint is not sufficient for the fractional terms to have a positive value. We do not treat

2.2 CCR Model

this assumption in explicit mathematical form at this time. Instead we put this in managerial terms by assuming that all outputs and inputs have some nonzero worth and this is to be reflected in the weights v and u being assigned some positive value.

Given the data, we measure the efficiency of each DMU once and hence need n optimizations, one for each DMU to be evaluated.

Definition 2.1 (CCR Efficiency). DMU_o is CCR-efficient if $\theta^* = 1$ and there exists at least one optimal $u^* > 0$ and $v^* > 0$.

We now replace the above fractional program (FP) by the following linear program (LP):

$$\begin{cases} \max_{u,v} \theta = v^T y_0 \\ \text{subject to:} \\ \quad u^T x_0 = 1 \\ \quad v^T y_j - u^T x_j \leq 0, \ j = 1, 2, \cdots, n \\ \quad u \geq 0 \\ \quad v \geq 0. \end{cases} \quad (2.2)$$

Theorem 2.1. *The fractional program (2.1) is equivalent to the linear program (2.2).*

The dual problem of the linear program (2.2) is expressed with a real variable θ and a nonnegative vector $\lambda = (\lambda_1, \lambda_2, \cdots, \lambda_n)$ of variables as follows:

$$\begin{cases} \theta = \min \theta \\ \text{subject to:} \\ \quad \sum_{j=1}^{n} x_{ij} \lambda_j \leq \theta x_{i0}, \quad i = 1, 2, \cdots, p \\ \quad \sum_{j=1}^{n} y_{rj} \lambda_j \geq y_{r0}, \quad r = 1, 2, \cdots, q \\ \quad \lambda_j \geq 0, \quad j = 1, 2, \cdots, n. \end{cases} \quad (2.3)$$

This model (2.3) is sometimes referred to as the "Farrell model" because it is the one used in Farrell. In the economics portion of the DEA literature, it is said to conform to the assumption of "strong disposal," but the efficiency evaluation it makes ignores the presence of nonzero slacks. In the operations research portion of the DEA literature, this is referred to as "weak efficiency."

The dual model (2.3) has a feasible solution $\theta^* = 1, \lambda_0^* = 1, \lambda_j^* = 0 \ (j \neq 0)$. Hence the optimal value θ^* is not greater than 1. The optimal solution, θ^*, yields an efficiency score for a particular DMU. The process is repeated for each DMU_j, $j = 1, 2, \cdots, n$. DMUs for which $\theta^* < 1$ are inefficient, while DMUs for which $\theta^* = 1$ are boundary points.

Some boundary points may be "weakly efficient" because we have nonzero slacks. This may appear to be worrisome because alternate optima may have nonzero slacks in some solutions, but not in others. However, we can avoid being worried even in such cases by invoking the following linear program in which the slacks are taken to their maximal values:

$$\begin{cases} \max \sum_{i=1}^{m} s_i^- + \sum_{r=1}^{s} s_r^+ \\ \text{subject to:} \\ \sum_{j=1}^{n} x_{ij}\lambda_j + s_i^- = \theta^* x_{i0}, \quad i = 1, 2, \ldots, p \\ \sum_{j=1}^{n} y_{rj}\lambda_j - s_r^+ = y_{r0}, \quad r = 1, 2, \ldots, q \\ \lambda_j \geq 0, \quad j = 1, 2, \cdots, n \\ s_i^- \geq 0, \quad i = 1, 2, \cdots, p \\ s_r^+ \geq 0, \quad r = 1, 2, \cdots, q \end{cases} \quad (2.4)$$

where we note the choices of s_i^- and s_r^+ do not affect the optimal θ^*, which is determined from model (2.3).

These developments now lead to the following definitions based upon the "relative efficiency" in Definition 2.1.

Definition 2.2 (DEA Efficiency). The performance of DMU_0 is fully (100%) efficient if and only if both (1) $\theta^* = 1$ and (2) all slacks $s_i^{-*} = s_r^{+*} = 0$.

Definition 2.3 (Weakly DEA Efficiency). The performance of DMU_0 is weakly efficient if and only if both (1) $\theta^* = 1$ and (2) $s_i^{-*} \neq 0$ and/or $s_r^{+*} \neq 0$ for some i or r in some alternate optima.

It is to be noted that the preceding development amounts to solving the following problem in two steps:

$$\begin{cases} \min \theta - \varepsilon \left(\sum_{i=1}^{m} s_i^- + \sum_{r=1}^{s} s_r^+ \right) \\ \text{subject to:} \\ \sum_{j=1}^{n} x_{ij}\lambda_j + s_i^- = \theta x_{i0}, \quad i = 1, 2, \cdots, p \\ \sum_{j=1}^{n} y_{rj}\lambda_j - s_r^+ = y_{r0}, \quad r = 1, 2, \cdots, q \\ \lambda_j \geq 0, \quad j = 1, 2, \cdots, n \\ s_i^- \geq 0, \quad i = 1, 2, \cdots, p \\ s_r^+ \geq 0, \quad r = 1, 2, \cdots, q \end{cases} \quad (2.5)$$

2.2 CCR Model

where the s_i^- and s_r^+ are slack variables used to convert the inequalities in (2.3) to equivalent equations. Here, $\varepsilon > 0$ is a so-called non-Archimedean element defined to be smaller than any positive real number. This is equivalent to solving (2.3) in two stages by first minimizing θ and then fixing $\theta = \theta^*$ as in (2.4), where the slacks are to be maximized without altering the previously determined value of $\theta = \theta^*$.

Alternately, one could have started with the output side and considered instead the ratio of virtual input to output. This would reorient the objective from max to min, as in (2.1), to obtain

$$\begin{cases} \max_{u,v} \theta = \dfrac{u^T x_0}{v^T y_0} \\ \text{subject to:} \\ u^T x_j \leq v^T y_j, \ j = 1, 2, \cdots, n \\ u \geq \varepsilon > 0 \\ v \geq \varepsilon > 0 \end{cases} \quad (2.6)$$

where $\varepsilon > 0$ is the previously defined non-Archimedean element.

Similar to model (2.2) and (2.5), the input models are as follows:

$$\begin{cases} \max_{u,v} \theta = u^T x_0 \\ \text{subject to:} \\ v^T y_0 = 1 \\ u^T x_j - v^T y_j \geq 0, \ j = 1, 2, \cdots, n \\ u \geq \varepsilon > 0 \\ v \geq \varepsilon > 0, \end{cases} \quad (2.7)$$

and

$$\begin{cases} \max \phi + \varepsilon \left(\sum_{i=1}^{m} s_i^- + \sum_{r=1}^{s} s_r^+ \right) \\ \text{subject to:} \\ \sum_{j=1}^{n} x_{ij} \lambda_j + s_i^- = x_{io}, \quad i = 1, 2, \cdots, p \\ \sum_{j=1}^{n} y_{rj} \lambda_j - s_r^+ = \phi y_{ro}, \quad r = 1, 2, \cdots, q \\ \lambda_j \geq 0, \quad j = 1, 2, \cdots, n \\ s_i^- \geq 0, \quad i = 1, 2, \cdots, p \\ s_r^+ \geq 0, \quad r = 1, 2, \cdots, q \end{cases} \quad (2.8)$$

See Cooper et al. [7] for a formal development of this transformation and modification of the expression for $\varepsilon > 0$.

Table 2.1 CCR DEA model

Envelopment model	Multiplier model
Input-oriented	
$\min \theta - \varepsilon \left(\sum_{i=1}^{m} s_i^- + \sum_{r=1}^{s} s_r^+ \right)$	$\max z = \sum_{r=1}^{q} \mu_r y_{r0}$
Subject to:	Subject to:
$\sum_{j=1}^{n} x_{ij} \lambda_j + s_i^- = \theta x_{i0} \quad i = 1, 2, \cdots, p$	$\sum_{r=1}^{q} \mu_r y_{rj} - \sum_{i=1}^{q} v_i y_{ij} \leq 0$
$\sum_{j=1}^{n} y_{rj} \lambda_j - s_r^+ = y_{r0} \quad r = 1, 2, \cdots, q$	$\sum_{i=1}^{p} v_i x_{i0} = 1$
$\lambda_j \geq 0 \quad j = 1, 2, \cdots, n$	$u_r, v_i \geq \varepsilon > 0$
Output-oriented	
$\max \phi + \varepsilon \left(\sum_{i=1}^{m} s_i^- + \sum_{r=1}^{s} s_r^+ \right)$	$\min q = \sum_{i=1}^{p} v_i x_{i0}$
Subject to:	Subject to:
$\sum_{j=1}^{n} x_{ij} \lambda_j + s_i^- = x_{i0} \quad i = 1, 2, \cdots, p$	$\sum_{i=1}^{p} v_i x_{ij} - \sum_{r=1}^{q} \mu_r y_{rj} \geq 0$
$\sum_{j=1}^{n} y_{rj} \lambda_j - s_r^+ = \phi y_{r0} \quad r = 1, 2, \cdots, q$	$\sum_{r=1}^{q} \mu_r y_{r0} = 1$
$\lambda_j \geq 0 \quad j = 1, 2, \cdots, n$	$\mu_r, v_i \geq \varepsilon > 0$

Here, we use a model with an output-oriented objective as contrasted with the input orientation in (1.6). However, as before, model (1.9) is calculated in a two-stage process. First, we calculate ϕ^* by ignoring the slacks. Then we optimize the slacks by fixing ϕ^* in the following linear programming problem:

$$\begin{cases} \max \sum_{i=1}^{p} s_i^- + \sum_{r=1}^{q} s_r^+ \\ \text{subject to:} \\ \quad \sum_{j=1}^{n} x_{ij} \lambda_j + s_i^- = x_{i0} \quad i = 1, 2, \cdots, p \\ \quad \sum_{j=1}^{n} y_{rj} \lambda_j - s_r^+ = \phi^* y_{r0} \quad r = 1, 2, \cdots, q \\ \quad \lambda_j \geq 0 \quad j = 1, 2, \cdots, n. \end{cases} \qquad (2.9)$$

Table 2.1 presents the CCR model in input-and-output-oriented versions, each in the form of a pair of dual linear programs.

2.3 BCC Model

The input-oriented BCC model proposed by Banker et al. [1] evaluates the efficiency of DMU$_0$ by solving the following linear program:

2.3 BCC Model

$$\begin{cases} \theta_B = \min \theta \\ \text{subject to:} \\ \quad \sum_{j=1}^{n} x_{ij}\lambda_j \leq \theta x_{i0}, \quad i = 1, 2, \cdots, p \\ \quad \sum_{j=1}^{n} y_{rj}\lambda_j \geq y_{r0}, \quad r = 1, 2, \cdots, q \\ \quad \sum_{k=1}^{n} \lambda_k = 1 \\ \quad \lambda_k \geq 0, \quad k = 1, 2, \cdots, n. \end{cases} \quad (2.10)$$

Some boundary points may be "weakly efficient" because we have nonzero slacks. This may appear to be worrisome because alternate optima may have nonzero slacks in some solutions, but not in others. However, we can avoid being worried even in such cases by invoking the following linear program in which the slacks are taken to their maximal values:

$$\begin{cases} \max \sum_{i=1}^{m} s_i^- + \sum_{r=1}^{s} s_r^+ \\ \text{subject to:} \\ \quad \sum_{j=1}^{n} x_{ij}\lambda_j + s_i^- = \theta^* x_{i0}, \quad i = 1, 2, \cdots, p \\ \quad \sum_{j=1}^{n} y_{rj}\lambda_j - s_r^+ = y_{r0}, \quad r = 1, 2, \cdots, q \\ \quad \sum_{k=1}^{n} \lambda_k = 1 \\ \quad \lambda_k \geq 0, \quad k = 1, 2, \cdots, n \\ \quad s_i^- \geq 0, \quad i = 1, 2, \cdots, p \\ \quad s_j^+ \geq 0, \quad j = 1, 2, \cdots, q. \end{cases} \quad (2.11)$$

It is to be noted that the preceding development amounts to solving the following problem in two steps:

$$\begin{cases} \min \theta - \varepsilon \left(\sum_{i=1}^{m} s_i^- + \sum_{r=1}^{s} s_r^+ \right) \\ \text{subject to:} \\ \quad \sum_{j=1}^{n} x_{ij}\lambda_j + s_i^- = \theta x_{i0}, \quad i = 1, 2, \cdots, p \\ \quad \sum_{j=1}^{n} y_{rj}\lambda_j - s_r^+ = y_{r0}, \quad r = 1, 2, \cdots, q \\ \quad \sum_{k=1}^{n} \lambda_k = 1 \\ \quad \lambda_k \geq 0, \quad k = 1, 2, \cdots, n \\ \quad s_i^- \geq 0, \quad i = 1, 2, \cdots, p \\ \quad s_j^+ \geq 0, \quad j = 1, 2, \cdots, q. \end{cases} \quad (2.12)$$

The dual multiplier form of the linear program (2.10) is expressed as

$$\begin{cases} \max_{u,v,v_0} \theta_B = v^T y_0 - v_0 \\ \text{subject to:} \\ \quad u^T x_0 = 1 \\ \quad v^T y_j - u^T x_j - v_0 \leq 0, \; j = 1, 2, \cdots, n \\ \quad u \geq 0 \\ \quad v \geq 0 \end{cases} \quad (2.13)$$

where v and u are vectors. The scalar v_0 may be positive or negative (or zero). The equivalent BCC fractional program is obtained from the dual program (2.13) as

$$\begin{cases} \max_{u,v} \theta_B = \dfrac{v^T y_0 - v_0}{u^T x_0} \\ \text{subject to:} \\ \quad \dfrac{v^T y_j - v_0}{u^T x_j} \leq 1, \; j = 1, 2, \cdots, n \\ \quad u \geq 0 \\ \quad v \geq 0. \end{cases} \quad (2.14)$$

It is clear that a difference between the CCR and BCC models is present in the free variable v_0, which is the dual variable associated with the constraint $\sum_{k=1}^{n} \lambda_k = 1$ that also does not appear in the CCR model. In the first phase, we minimize θ by model (2.10) and, in the second phase, we maximize the sum of the input excesses and output shortfalls, keeping θ^* (the optimal objective value obtained in Phase one) by model (2.11). The evaluations secured from the CCR and BCC models are also related to each other as follows. An optimal solution for (2.10) and (2.11) is represented by $(\theta_B^*, s^{-*}, s^{+*})$, where s^{-*} and s^{+*} represent the maximal input excesses and output shortfalls, respectively. Notice that θ_B^* is not less than the optimal objective value θ^* of the CCR model, since (2.10) imposes one additional constraint, $\sum_{k=1}^{n} \lambda_k = 1$, so its feasible region is a subset of feasible region for the CCR model.

Definition 2.4 (BCC Efficiency). If an optimal solution $(\theta_B^*, s^{-*}, s^{+*})$ obtained in this two-phase process for model (2.10) satisfies $\theta_B^* = 1$ and has no slack $s^{-*} = s^{+*} = 0$, then the DMU$_0$ is called BCC-efficient; otherwise it is BCC-inefficient.

Theorem 2.2. *The improved activity* $(\theta^* x - s^{-*}, y + s^{+*})$ *is BCC-efficient.*

Theorem 2.3. *A DMU that has a minimum input value for any input item, or a maximum output value for any output item, is BCC-efficient.*

The output-oriented BCC model is

$$\begin{cases} \max \ \eta \\ \text{subject to:} \\ \quad \sum_{j=1}^{n} x_{ij}\lambda_j \leq x_{i0}, \quad i = 1, 2, \cdots, p \\ \quad \sum_{j=1}^{n} y_{rj}\lambda_j \geq \eta y_{r0}, \quad r = 1, 2, \cdots, q \\ \quad \sum_{k=1}^{n} \lambda_k = 1 \\ \quad \lambda_k \geq 0, \quad k = 1, 2, \cdots, n. \end{cases} \quad (2.15)$$

The dual form associated with the above linear program (2.15) is expressed as

$$\begin{cases} \min_{\boldsymbol{u},\boldsymbol{v},u_0} \ \boldsymbol{v}^T \boldsymbol{y}_0 - v_0 \\ \text{subject to:} \\ \quad \boldsymbol{v}^T \boldsymbol{y}_0 = 1 \\ \quad \boldsymbol{u}^T \boldsymbol{x}_j - \boldsymbol{v}^T \boldsymbol{y}_j - v_0 \geq 0, \ j = 1, 2, \cdots, n \\ \quad \boldsymbol{u} \geq 0 \\ \quad \boldsymbol{v} \geq 0 \end{cases} \quad (2.16)$$

where v_0 is the scalar associated with $\sum_{k=1}^{n} \lambda_k = 1$ in the envelopment model. Finally, we have the equivalent (BCC) fractional programming formulation for model (2.16):

$$\begin{cases} \min_{\boldsymbol{u},\boldsymbol{v},u_0} \ \dfrac{\boldsymbol{u}^T \boldsymbol{x}_0 - u_0}{\boldsymbol{v}^T \boldsymbol{y}_0} \\ \text{subject to:} \\ \quad \dfrac{\boldsymbol{u}^T \boldsymbol{x}_j - v_0}{\boldsymbol{v}^T \boldsymbol{y}_j} \geq 1, \ j = 1, 2, \cdots, n \\ \quad \boldsymbol{u} \geq 0 \\ \quad \boldsymbol{v} \geq 0. \end{cases} \quad (2.17)$$

2.4 Additive Model

The preceding models required us to distinguish between input-oriented and output-oriented models. Now, however, we combine both orientations in a single model, called additive model proposed by Charnes et al. [3].

Let us consider a production possibility set (PPS), consisting of all convex combinations of $(x_k, y_k), k = 1, 2, \cdots, n$. We can formulate it as

$$\text{PPS} = \left\{ (x, y) \middle| x = \sum_{k=1}^{n} x_k \lambda_k, \ y = \sum_{k=1}^{n} y_k \lambda_k, \ \sum_{k=1}^{n} \lambda_k = 1, \ \lambda_1 \geq 0, \ \lambda_2 \geq 0, \cdots, \lambda_n \geq 0 \right\}.$$

Then the additive model can be given as

$$\begin{cases} \max \ \sum_{i=1}^{p} s_i^- + \sum_{j=1}^{q} s_j^+ \\ \text{subject to:} \\ \quad \sum_{k=1}^{n} x_{ki} \lambda_k = x_{0i} - s_i^-, \quad i = 1, 2 \cdots, p \\ \quad \sum_{k=1}^{n} y_{kj} \lambda_k = y_{0j} + s_j^+, \quad j = 1, 2, \cdots, q \\ \quad \sum_{k=1}^{n} \lambda_k = 1 \\ \quad \lambda_k \geq 0, \quad k = 1, 2, \cdots, n \\ \quad s_i^- \geq 0, \quad i = 1, 2, \cdots, p \\ \quad s_j^+ \geq 0, \quad j = 1, 2, \cdots, q \end{cases} \quad (2.18)$$

where s_i^- and s_j^+ represent output and input slacks, respectively.

It is clear that this model considers the total slacks of inputs and output simultaneously in arriving at a point on the efficient frontier.

Definition 2.5 (ADD Efficiency). DMU_0 is ADD-efficient if s_i^{-*} and s_j^{+*} are zero for $i = 1, 2, \cdots, p$ and $j = 1, 2, \cdots, q$, where s_i^{-*} and s_j^{+*} are optimal solutions of (2.18).

DMU_0 is ADD-efficient if there is no $(x, y) \in$ PPS such that $x \leq x_0$ and $y \geq y_0$ with strict inequality holding for at least one of the components in the input or the output vector.

The dual problem to the additive model (2.18) can be expressed as follows:

$$\begin{cases} \max_{u, v, v_0} \ u^T x_0 - v^T y_0 + v_0 \\ \text{subject to:} \\ \quad v^T y_j - u^T x_j - v_0 \leq 0, \quad j = 1, 2, \cdots, n \\ \quad u \geq e \\ \quad v \geq e. \end{cases} \quad (2.19)$$

Theorem 2.4. *DMU_0 is ADD-efficient if and only if it is BCC-efficient.*

Theorem 2.5 (Tone [14]). *DMU_0 is CCR-efficient if and only if it is SBM-efficient.*

2.4 Additive Model

The model (2.18) uses a metric that differs from the one used in the "radial measure" model which uses what is called the ℓ_1 metric in mathematics, and the "city block metric" in operations research. It also dispenses with the need for distinguishing between an "output" and an "input" orientation as was done in the discussion leading up to (2.9) because the objective in (2.18) simultaneously maximizes outputs and minimizes inputs in the sense of vector optimizations. This can be seen by utilizing the solution to (2.18) to introduce new variables $\hat{y}_{ro}, \hat{x}_{io}$ defined as follows:

$$\hat{y}_{r0} = y_{r0} + s_r^{+*} \geq y_{r0}, r = 1, \cdots, q$$
$$\hat{x}_{i0} = x_{i0} - s_i^{-*} \leq x_{i0}, i = 1, \cdots, p. \qquad (2.20)$$

Now, note that the slacks are all independent of each other. Hence, an optimum is not reached until it is not possible to increase an output $\hat{y}_{ro}$ or reduce an input $\hat{x}_{io}$ without decreasing some other output or increasing some other input.

We now use the class of additive models to develop a different route to treating technical, allocative, and overall inefficiencies and their relations to each other. This can help to avoid difficulties in treating possibilities such as negative or zero profits, which are not easily treated by the ratio approaches, which are commonly used in the DEA literature. See the discussion in Cooper et al. [4,5] from which the following development is taken. See also Chap. 8 in Cooper et al. [7].

First, we observe that we can multiply the output slacks by unit prices and the input slacks by unit costs after we have solved (1.19) and thereby accord a monetary value to this solution. Then, we can utilize (1.20) to write

$$\sum_{r=1}^{s} p_{ro} s_r^{+*} + \sum_{i=1}^{m} c_{io} s_i^{-*}$$
$$= \left(\sum_{r=1}^{s} p_{ro} \hat{y}_{ro} - \sum_{r=1}^{s} p_{ro} y_{ro} \right) + \left(\sum_{i=1}^{m} c_{io} x_{io} - \sum_{i=1}^{m} c_{io} \hat{x}_{io} \right)$$
$$= \left(\sum_{r=1}^{s} p_{ro} \hat{y}_{ro} - \sum_{i=1}^{m} c_{io} \hat{x}_{io} \right) - \left(\sum_{r=1}^{s} p_{ro} y_{ro} - \sum_{i=1}^{m} c_{io} x_{io} \right). \qquad (2.21)$$

From the last pair of parenthesized expressions, we find that, at an optimum, the objective in (2.18) after multiplication by unit prices and costs is equal to the profit available when production is technically efficient minus the profit obtained from the observed performance. Hence, when multiplied by unit prices and costs, the solution to (2.18) provides a measure in the form of the amount of the profits lost by not performing in a technically efficient manner term by term if desired.

We can similarly develop a measure of allocative efficiency by means of the following additive model:

$$\begin{cases} \max \sum_{r=1}^{s} p_{r0}\hat{s}_r^+ + \sum_{i=1}^{m} c_{i0}\hat{s}_i^- \\ \text{subject to:} \\ \quad \hat{y}_{r0} = \sum_{j=1}^{n} y_{rj}\hat{\lambda}_j - \hat{s}_r^+, \, r = 1, 2, \cdots, q \\ \quad \hat{x}_{i0} = \sum_{j=1}^{n} x_{ij}\hat{\lambda}_j + \hat{s}_i^-, \, i = 1, 2, \cdots, p \\ \quad \sum_{j=1}^{n} \hat{\lambda}_j = 1 \\ \quad \hat{\lambda}_j \geq 0, \quad j = 1, 2, \cdots, n. \end{cases} \quad (2.22)$$

2.5 SBM Model

We now augment the additive models by introducing a measure that makes its efficiency evaluation, as effected in the objective, invariant to the units of measure used for the different inputs and outputs. That is, we would like this summary measure to assume the form of a scalar that yields the same efficiency value when distances are measured in either kilometers or miles. More generally, we want this measure to be the same when x_{ij} are replaced by $k_i x_{ij}$ and y_{rj} are replaced by $c_r y_{rj}$, where the k_i and c_r are arbitrary positive constants, $i = 1, 2 \cdots, p$, $j = 1, 2 \cdots, q$.

This property is known by names such as "dimension free" (see [13]) and "units invariant." In this section, we introduce such a measure for additive models in the form of a single scalar called "SBM" (Slacks-Based Measure), which was introduced by Tone [14] and has the following important properties:

P1 The measure is invariant with respect to the unit of measurement of each input and output item. (Units invariant)
P2 The measure is monotone decreasing in each input and output slack. (Monotone)

In order to estimate the efficiency, we formulate the following fractional program:

$$\begin{cases} \min \quad \rho = \dfrac{1 - \frac{1}{p}\sum_{i=1}^{p} s_i^-/x_{0i}}{1 + \frac{1}{q}\sum_{j=1}^{q} s_j^+/y_{0j}} \\ \text{subject to:} \\ \quad \sum_{k=1}^{n} x_{ki}\lambda_k = x_{0i} - s_i^-, \quad i = 1, 2 \cdots, p \\ \quad \sum_{k=1}^{n} y_{kj}\lambda_k = y_{0j} + s_j^+, \quad j = 1, 2, \cdots, q \\ \quad \lambda_k \geq 0, \quad k = 1, 2, \cdots, n \\ \quad s_i^- \geq 0, \quad i = 1, 2, \cdots, p \\ \quad s_j^+ \geq 0, \quad j = 1, 2, \cdots, q. \end{cases} \quad (2.23)$$

In this model, we assume that $x_k \geq 0$, $k = 1, 2, \cdots, n$. If $x_{0i} = 0$, we delete the term s_i^-/x_{0i} in the objective function. If $y_{0j} \leq 0$, we replace it by a very small positive number so that the term s_i^-/x_{0i} plays a role of penalty.

It is readily verified that the objective function value ρ satisfies (PI) because the numerator and denominator are measured in the same units for every item in the objective of (2.23). It is also readily verified that an increase in either s_i^- or s_j^+, all else held constant, will decrease this objective value and, indeed, do so in a strictly monotone manner.

Furthermore, we have

$$0 \leq \rho \leq 1.$$

The formula for ρ in (2.23) can be transformed into

$$\rho = \left(\frac{1}{p} \sum_{i=1}^p \frac{x_{0i} - s_i^-}{x_{0i}} \right) \left(\frac{1}{q} \sum_{i=1}^q \frac{y_{0j} + s_j^+}{y_{0j}} \right)^{-1}. \qquad (2.24)$$

The ratio $(x_{0i} - s_i^-)/x_{0i}$ evaluates the relative reduction rate of input i and, therefore, the first term corresponds to the mean proportional reduction rate of inputs or input mix inefficiencies. Similarly, in the second term, the ratio $(y_{0j} + s_j^+)/y_{0j}$ evaluates the relative proportional expansion rate of output j and $(1/q) \sum_{i=1}^q (y_{0j} + s_j^+)/y_{0j}$ is the mean proportional rate of output expansion. Its inverse, the second term, measures output mix inefficiency. Thus, ρ can be interpreted as the ratio of mean input and output mix inefficiencies. Further, we have the following theorem.

Theorem 2.6. *If DMU_A dominates DMU_B so that $x_A \leq x_B$ and $y_A \geq y_B$, then $\rho_A^* \geq \rho_B^*$.*

Definition 2.6 (SBM Efficiency). DMU_0 is SBM-efficient if and only if $\rho^* = 1$, where $\rho^* = 1$ is the optimal value of (2.23).

Theorem 2.7. *The optimal ρ^* in SBM model (2.23) is not greater than the optimal θ^* in CCR model (2.1).*

2.6 Russell Measure Model

We now introduce a model described as the "Russell Measure Model." Actually it was introduced and named by Färe and Lovell [8]. Their formulation is difficult to compute, however, so we turn to a more recent development due to Pastor et al. [11]. This model is given as follows:

$$\begin{cases} \psi = \min\limits_{\theta,\eta} \dfrac{\sum_{i=1}^{p} \theta_i/p}{\sum_{j=1}^{q} \eta_j/q} \\ \text{subject to:} \\ \quad \sum\limits_{k=1}^{n} x_{ki}\lambda_k \leq \theta_i x_{0i}, \quad i=1,2\cdots,p \\ \quad \sum\limits_{k=1}^{n} y_{kj}\lambda_k \geq \eta_j y_{0j}, \quad j=1,2,\cdots,q \\ \quad \lambda_k \geq 0, \quad k=1,2,\cdots,n \\ \quad 0 \leq \theta_i \leq 1, \quad i=1,2,\cdots,p \\ \quad \eta_j \geq 1, \quad j=1,2,\cdots,q. \end{cases} \quad (2.25)$$

Pastor et al. [11] refer to this as the "Enhanced Russell Graph Measure of Efficiency," but we shall refer to it as ERM (Enhanced Russell Measure). See Färe et al. [9] for the meaning of "graph measure." Such measures are said to be "closed," so ψ includes all inefficiencies that the model can identify. In this way we avoid limitations of the radial measures which cover only some of the input or output inefficiencies and hence measure only "weak efficiency."

The closure property is shared by SBM. In fact SBM and ERM are related as in the following theorem.

Theorem 2.8. *ERM as formulated in (2.25) and SBM as formulated in (2.23) are equivalent in that λ values that are optimal for one are also optimal for the other.*

References

1. Banker RD, Charnes A, Cooper WW (1984) Some models for estimating technical and scale efficiencies in data envelopment analysis. Manag Sci 30:1078–1092
2. Charnes A, Cooper WW, Rhodes E (1978) Measuring the efficiency of decision making units. Eur J Oper Res 2:429–444
3. Charnes A, Cooper WW, Golany B, Seiford L, Stutz J (1985) Foundations of data envelopment analysis for Pareto-Koopmans efficient empirical production functions. J Econom 30:91–107
4. Cooper WW, Park KS, Yu G (1999) IDEA and AR-IDEA: models for dealing with imprecise data in DEA. Manag Sci 45:597–607
5. Cooper WW, Park KS, Pastor JT (1999) RAM: a range adjusted measure of inefficiency for use with additive models, and relations to other models and measures in DEA. J Product Anal 11:5–24
6. Cooper WW, Seiford LM, Tone K (2000) Data envelopment analysis: a comprehensive text with models, applications, references, and DEA-Solver software. Kluwer Academic, Boston
7. Cooper WW, Seiford LM, Tone K (2007) Data envelopment analysis: a comprehensive text with models, applications, references and DEA-solver software. Kluwer, Boston
8. Färe R, Lovell CAK (1978) Measuring the technical efficiency of production. J Econ Theory 19:150–162
9. Färe R, Grosskopf S, Lovell CAK (1985) The measurement of efficiency of production. Kluwer-Nijhoff, Boston
10. Farrell MJ (1957) The measurement of productive efficiency. J R Stat Soc A 120:253–281

11. Pastor JT, Ruiz JL, Sirvent I (1999) An enhanced DEA Russell graph efficiency measure. Eur J Oper Res 115:596–607
12. Petersen NC (1990) Data envelopment analysis on a relaxed set of assumptions. Manag Sci 36(3):305–313
13. Thrall RM (1996) Duality, classification and slacks in DEA. Ann Oper Res 66:23–125
14. Tone K (2001) A slack-based measure of efficiency in data envelopment analysis. Eur J Oper Res 130:498–509

Chapter 3
Stochastic DEA

Although DEA offers more advantages than many other statistical approaches, some limitations have to be considered. One important problem is its sensitivity to data. Therefore, a key to the success of the DEA approach is the accurate measure of all factors, including that of inputs and outputs. However, in many situations, such as in a manufacturing system, in a production process, or in a service system, inputs and outputs are so volatile and complex that they are difficult to measure in an accurate way. Thus, some researchers have proposed several models to deal with the data variation in DEA by stochastic models. Sengupta [32] generalized the stochastic DEA model by using the expected value to the stochastic inputs and outputs. Banker [3] incorporated statistical elements into DEA and developed an approach which aims to effect inferences in statistical noise. Many papers (Olesen and Petersen [30], Banker [2], Cooper [12, 14], Land [25]) have introduced chance-constrained programming to DEA in order to accommodate stochastic variations in data. Additional stochastic DEA approaches can be found in Horace [21], Gong [19], Simar [33, 34], and Grosskopf [20].

This chapter deals with chance-constrained programming extensions of the usual deterministic DEA formulations. This kind of approach makes it possible to replace deterministic characterizations in DEA, such as replacing "efficient" and "not efficient" with such characterizations as "probably efficient" and "probably not efficient." Indeed, it is possible to go still further into characterizations such as "sufficiently efficient" with associated probabilities of being not correct in making inferences about the performance of a decision-making unit (DMU).

It is also possible to extend the deterministic objectives usually used in DEA with additional alternatives. For instance, one may use the "E-model" of chance-constrained programming to obtain an "expected value" approach. However, this expected value objective is not the only possibility. One may also use the "P-model" of chance-constrained programming to obtain the "most probable" occurrences, perhaps in order to determine whether this probability is sufficiently high. Indeed, one can extend this by incorporating constraints (also probabilistic in character) to ensure that the resulting solutions are satisfactory.

These are the ideas and extensions that will be covered in this chapter. The purpose of this chapter, however, is to provide a systematic presentation of major developments of chance-constrained DEA models that have been discussed in the literature.

3.1 Symbols and Notations

Suppose there are n DMUs and the symbols and notations are listed as follows:

DMU_i: the ith DMU, $i = 1, 2, \cdots, n$
DMU_0: the target DMU
$\tilde{x}_i = (\tilde{x}_{1i}, \cdots, \tilde{x}_{mi})^T$: the "random input" vector of DMU_i, $i = 1, 2, \cdots, n$, respectively
$\tilde{y}_i = (\tilde{y}_{1i}, \cdots, \tilde{y}_{ri})^T$: the "random output" vector of DMU_i, $i = 1, 2, \cdots, n$, respectively
$\tilde{x}_0 = (\tilde{x}_{10}, \cdots, \tilde{x}_{m0})^T$: the "random input" vector of DMU_0
$\tilde{y}_0 = (\tilde{y}_{10}, \cdots, \tilde{y}_{r0})^T$: the "random output" vector of DMU_0
$\tilde{Y} = (\tilde{y}_1, \cdots, \tilde{y}_n)$: the $(s \times n)$ "output" matrix
$Y = (y_1, \cdots, y_n)$: the $(s \times n)$ "expected output" matrix
$_k\tilde{y} = (\tilde{y}_{k1}, \cdots, \tilde{y}_{kn})$: the kth row of the "output" matrix $\tilde{Y}$, $k = 1, \cdots, s$
$_ky = (y_{k1}, \cdots, y_{kn})$: the kth row of the "expected output" matrix Y, $k = 1, \cdots, s$
$\tilde{X} = (\tilde{x}_1, \cdots, \tilde{x}_n)$: the $(m \times n)$ "input" matrix
$X = (x_1, \cdots, x_n)$: the $(m \times n)$ "expected input" matrix
$_i\tilde{x} = (\tilde{x}_{i1}, \cdots, \tilde{x}_{in})$: the ith row of the "input" matrix $\tilde{X}$, $i = 1, \cdots, m$
$_ix = (x_{i1}, \cdots, x_{in})$: the ith row of the "expected input" matrix X, $i = 1, \cdots, m$

The probability distributions of $\tilde{x}_{ij}$ and $\tilde{y}_{rj}$ will usually be determined by historical data on the inputs and outputs, but we may replace some or all of these historically determined probability distributions by theoretical probability distributions, as we shall do when this serves our purposes.

3.2 Stochastic DEA Models

Using the notations in Sect. 3.1, we can extend our characterizations to "stochastic efficiency dominance" which can be obtained from the joint probabilistic comparisons of its outputs and inputs with every other observed DMU. Thus, informally, if $\tilde{y}_0$ and $\tilde{x}_0$ are the output and input vectors of the DMU_0 to be tested relative to all DMU_j, $j = 1, \cdots, n$, we will say that DMU_0 is stochastically not dominated in its efficiency if it is stochastically impossible to augment any output without increasing any input and without decreasing any other output or if it is stochastically impossible to decrease any input without augmenting any other input and without decreasing any output. This is intended as a stochastic generalization of efficiency

3.2 Stochastic DEA Models

dominance as defined in the preceding section. It is also an adaptation of Pareto-Koopmans efficiency to stochastic situations with the discrete production set $\tilde{T}_0 = \{(\tilde{x}_j, \tilde{y}_j)\}_{1 \le j \le n}$. This direct generalization to stochastic situations could be very restricted because of random variations in inputs and outputs. Therefore, we could incorporate a tolerance or risk level to the definition. For a given scalar α ($0 \le \alpha < 1$), DMU_0 is not stochastically dominated in its efficiency if and only if there is a joint probability less than or equal to a that some other observed DMU displays efficiency dominance relative to DMU_0.

Definition 3.1 (Cooper et al. [14]). DMU_0 is not stochastically dominated in its efficiency with respect to $\tilde{T}_0$ if and only if for all λ satisfying $e^T \lambda = 1, \lambda_j \in \{0, 1\}$, i.e., the components of λ are bivalent, we have

$$\Pr\left\{\bigcap_{i=1}^{m}(_i\tilde{x}\lambda \le \tilde{x}_{i0}) \bigcap_{r=1}^{s}(_r\tilde{y}\lambda \ge \tilde{y}_{r0})\right\}$$

$$= \Pr\left\{\sum_{j=1}^{n}\lambda_j\tilde{x}_{ij} \le \tilde{x}_{i0}, \sum_{j=1}^{n}\lambda_j\tilde{y}_{rj} \ge \tilde{y}_{r0}, i = 1, \cdots, m, r = 1, \cdots, s\right\}$$

$$\le \alpha. \tag{3.1}$$

Note that our definition can be applied to any probability distribution of inputs and outputs for the DMUs to be considered. Also note that if $\tilde{y}_j$ and $\tilde{x}_j$ follow a continuous joint probability distribution, the requirement of at least one strict inequality in the above definition is not necessary.

The model in (2.18) is deterministic, so we now provide a stochastic alternative via the following formulation:

$$\max \quad \Pr\left\{\bigcap_{i=1}^{m}(_i\tilde{x}\lambda \le \tilde{x}_{i0}) \bigcap_{r=1}^{s}(_r\tilde{y}\lambda \ge \tilde{y}_{r0})\right\} = \eta. \tag{3.2}$$

The formulations for stochastic efficiency dominance used above are based on the discrete stochastic production set

$$\tilde{T}_0 = \{(\tilde{x}, \tilde{y}) : \tilde{x} = \tilde{X}\lambda, \tilde{y} = \tilde{Y}\lambda, e^T\lambda = 1, \lambda_j \in \{0, 1\}, j = 1, \cdots, n\}$$

$$= \{(\tilde{x}_j, \tilde{y}_j)\}_{1 \le j \le n} \tag{3.3}$$

where e is a $(n \times 1)$ vector with all elements equal to unity. We extend this to the continuous stochastic production set $\tilde{T}$, in which the bivalency conditions on the variables λ_j are relaxed. These variables are now allowed to be continuous so that the required evaluation can be effected in terms of convex combinations of observed DMUs. Therefore, $\tilde{T}$ can be written as

$$\tilde{T} = \{(\tilde{x}, \tilde{y}) : \tilde{x} = \tilde{X}\lambda, \tilde{y} = \tilde{Y}\lambda, e^T\lambda = 1, \lambda \ge 0\}. \tag{3.4}$$

One of the associated stochastic production possibility sets of $\tilde{T}$ which is also considered here can be defined as

$$\tilde{T}_1 = \{(\tilde{x}, \tilde{y}) : \tilde{x} = \tilde{X}\lambda + s^+, \tilde{y} = \tilde{Y}\lambda - s^-, e^T\lambda = 1, \lambda \geq 0, s^+ \geq 0, s^- \geq 0\}. \tag{3.5}$$

We now note that this brings us into contact with other parts of the DEA literature. $\tilde{T}$ and $\tilde{T}_1$ are stochastic generalizations of the production possibility sets defined in Charnes et al. [8] and Banker et al. [4], respectively, where the additive and BCC models of DEA were first introduced into the literature. To see this, let us represent the general stochastic production possibility set as follows:

$$\tilde{T}_2 = \{(\tilde{x}, \tilde{y}) : \tilde{y} \text{ can be produced from } \tilde{x}\}. \tag{3.6}$$

We next postulate the following properties for the production possibility set (Cooper et al. [14]):

Postulate 1 (*Convexity*). If $(\tilde{x}_j, \tilde{y}_j) \in \tilde{T}_2$, $j = 1, \cdots, n$, and $\lambda_j \geq 0$ are non-negative scalars such that $e^T\lambda = 1$, then $(\tilde{X}\lambda, \tilde{Y}\lambda) \in \tilde{T}_2$.

Postulate 2 (*Inefficiency Postulate*). (a) If $(\tilde{x}, \tilde{y}) \in \tilde{T}_2$ and $\tilde{x}^* = \tilde{x} + s^+$ with $s^+ \geq 0$, then $(\tilde{x}^*, \tilde{y}) \in \tilde{T}_2$. (b) If $(\tilde{x}, \tilde{y}) \in \tilde{T}_2$ and $\tilde{y}^* = \tilde{y} - s^-$ with $s^- \geq 0$, then $(\tilde{x}, \tilde{y}^*) \in \tilde{T}_2$.

Postulate 3 (*Minimum Intersection*). $\tilde{T}_2$ is the intersection set of all $\hat{T}$ satisfying Postulates 1 and 2 and subject to the condition that each of the vectors $(\tilde{x}_j, \tilde{y}_j) \in \hat{T}$, $j = 1, \cdots, n$.

$$\tilde{T}_1 = \{(\tilde{x}, \tilde{y}) : \tilde{x} = \tilde{X}\lambda + s^+, \tilde{y} = \tilde{Y}\lambda - s^-, e^T\lambda = 1, \lambda \geq 0, s^+ \geq 0, s^- \geq 0\}$$

is thus a stochastic production possibility set satisfying the above three postulates. Furthermore, if we omit the convexity condition $e^T\lambda = 1$ in $\tilde{T}_1$, the production possibility set becomes a stochastic generalization of the production possibility set for the CCR model introduced in Charnes et al. [7]. Cooper et al. [14] have shown that both $\tilde{T}$ and $\tilde{T}_1$ have the same efficiency properties. Therefore, here we only discuss some major results on $\tilde{T}$.

To represent this explicitly, we introduce the "almost 100 % confidence" chance-constrained problem:

$$\begin{cases} \max \ \Pr\{e^T(\tilde{X}\lambda - \tilde{x}_0) + e^T(\tilde{y}_0 - \tilde{Y}\lambda) < 0\} \\ \text{subject to:} \\ \quad \Pr\{_i\tilde{x}\lambda < \tilde{x}_{i0}\} \geq 1 - \varepsilon, i = 1, \cdots, m \\ \quad \Pr\{_k\tilde{y}\lambda > \tilde{y}_{k0}\} \geq 1 - \varepsilon, k = 1, \cdots, s \\ \quad e^T\lambda = 1 \\ \quad \lambda \geq 0. \end{cases} \tag{3.7}$$

3.2 Stochastic DEA Models

Theorem 3.1 (Cooper et al. [14]). *From the above discussions, we then have the following:*

1. *Let DMU_0 be α-stochastically efficient. Then the optimal objective value of the "almost 100% confidence" chance-constrained programming problem (3.7) is less than or equal to α.*
2. *If the optimal objective value of (3.7) exceeds α, then DMU_0 is not stochastically efficient.*

The assumption of multivariate normality implies that the production possibility set and its efficient frontier will vary randomly in a symmetric manner across DMUs, and in this manner reflects the results of events such as bad weather and poor luck, etc., and it also permits data measurement and other errors to occur symmetrically.

This interpretation is similar to the two-sided error assumptions used by Aigner et al. [1] for the estimation of single output parametric stochastic frontier production functions. There is another one-sided error in their work, which represents a component that reflects an assumption that each DMUs output must lie on or below the stochastic frontier function if it is to represent inefficiency. Although we do not consider this one-sided disturbance explicitly in our stochastic DEA model, we do need to note that the structure of our stochastic production possibility set implicitly allows for one-sided disturbances from the efficient frontier due to possible DMU inefficiencies and this is reflected in our chance constraints being oriented in the direction where inefficiencies might occur. We restrict our consideration to the class of "zero-order decision rules" in chance-constrained programming to achieve a deterministic equivalent for the problem (3.7) as follows:

$$\begin{cases} \min \ e^T(X\lambda - x_0) + e^T(y_0 - Y\lambda) + \sigma(\lambda)\Phi^{-1}(\alpha) \\ \text{subject to:} \\ \quad y_{k0} \leq {}_k y\lambda + \sigma_k^O(\lambda)\phi^{-1}(\varepsilon), k = 1, \cdots, s \\ \quad {}_i x\lambda \leq x_{i0} + \sigma_i^I(\lambda)\phi^{-1}(\varepsilon), i = 1, \cdots, m \\ \quad e^T\lambda = 1 \\ \quad \lambda \geq 0 \end{cases} \quad (3.8)$$

for which we have the following theorem.

Theorem 3.2 (Cooper et al. [14]). *The model (3.8) has the following results:*

1. *Let DMU_0 be α-stochastically efficient. Then the optimal value of problem (3.8) is greater than or equal to zero.*
2. *If the optimal value of (3.8) is less than zero, then DMU_0 is not α-stochastically efficient.*

3.2.1 Marginal Chance-Constrained Models

Land et al. [25] introduced a formal "E-model" form of marginal chance-constrained DEA model in CCR (Charnes et al. [7]) form as follows:

$$\begin{cases} \min \ \theta \\ \text{subject to:} \\ \quad \Pr\left\{\theta \tilde{x}_{i0} \geq \sum_{j=1}^{n} \tilde{x}_{ij} \lambda\right\} \geq 1-\alpha, i=1,\cdots,m \\ \quad \Pr\left\{\sum_{j=1}^{n} \tilde{y}_{rj} \lambda_j \geq \tilde{y}_{r0}\right\} \geq 1-\alpha, r=1,\cdots,s \\ \quad \lambda_j \geq 0, j=1,\cdots,n. \end{cases} \quad (3.9)$$

The meaning of the chance constraints is that they should not be violated with probability at most a.

Olesen and Petersen [30] also utilized marginal chance-constrained programming theory to develop an "E-model" for use in DEA by introducing confidence regions for all DMUs. For given probability level γ,

$$D_j(\gamma) = \left\{(x,y) : \left(x^T - E(\tilde{x}_j)^T, y^T - E(\tilde{y}_j)^T\right) \sum\nolimits_j^{-1} \begin{pmatrix} x - E(\tilde{x}_j) \\ y - E(\tilde{y}_j) \end{pmatrix} \leq c^2\right\}$$

is called the confidence region of DMU$_j$, where $\sum_j$ is the covariance matrix of $(\tilde{x}_j, \tilde{y}_j)$, c is determined by $P\left(\chi^2_{(n)} \leq c^2\right) = \gamma$, and $\chi^2_{(n)}$ is the Chi-square random variable with n degrees of freedom. A comparison of these two types of approaches can be found in Olesen [29].

Letting $\alpha = 1 - \Phi(c)$, the chance-constrained DEA model is

$$\begin{cases} \max \ u^T y_0 - v^T x_0 \\ \text{subject to:} \\ \quad \Pr\left(u^T y_j \leq v^T x_j\right) \geq 1-\alpha, j=1,\cdots,n \\ \quad u \geq \varepsilon e, \\ \quad v \geq \varepsilon e \end{cases} \quad (3.10)$$

where ε is the non-Archimedean positive infinitesimal defined scalar and e is a vector of ones.

Differences between models in (3.9) and (3.10) are as follows:

1. The model in (3.9) generalized the CCR envelopment form to marginal chance-constrained formulations, while the model in (3.10) extended CCR multiplier models to marginal chance-constrained formulations.

3.2 Stochastic DEA Models

2. The scalar α was predetermined directly by the user in model (3.9), but in model (3.10) the scalar α was determined by another scalar γ through confidence regions of DMUs.

We here consider another version of E-model form for marginal chance-constrained DEA models (Cooper et al. [16–18]), which we will use to extend the concept of BCC efficiency to a chance-constrained programming context:

$$\begin{cases} \max \varphi \\ \text{subject to:} \\ \quad \Pr\left\{\sum_{j=1}^{n} \tilde{y}_{rj}\lambda_j \geq \varphi \tilde{y}_{r0}\right\} \geq 1-\alpha, \quad r=1,\cdots,s \\ \quad \Pr\left\{\sum_{j=1}^{n} \tilde{x}_{ij}\lambda_j \leq \tilde{x}_{i0}\right\} \geq 1-\alpha, \quad i=1,\cdots,m \\ \quad \sum_{j=1}^{n} \lambda_j = 1 \\ \quad \lambda_j \geq 0, \quad j=1,\cdots,n \end{cases} \quad (3.11)$$

in which $\alpha \in [0, 1]$ is a predetermined number.

Definition 3.2 (Chance-Constrained Efficiency). DMU$_0$ is stochastic efficient if and only if the following two conditions are both satisfied:

1. $\varphi^* = 1$.
2. Slack values are all zero for all optimal solutions.

Here (2) refers to all alternate optima because the second-stage optimization associated with $\varepsilon > 0$ is not used in (3.11).

Since $j = 0$ is one of the n DMU$_j$, we can always get a solution with $\phi = 1$, $\lambda_0 = 1$, and $\lambda_j = 0 (j \neq 0)$ and all slacks zero. However, this solution need not be maximal. It follows that a maximum with $\phi^* > 1$ in (3.11) for any sample of $j = 1, \cdots, n$ observations means that the DMU$_0$ being evaluated is not efficient because, to the specified level of probability defined by α, the evidence will then show that all outputs of DMU$_0$ can be increased to $\varphi^* \tilde{y}_{r0} > \tilde{y}_{r0}, r = 1, \cdots, s$, by using a convex combination of other DMUs which also satisfy

$$\Pr\left\{\sum_{j=1}^{n} \tilde{x}_{ij}\lambda_j \leq \tilde{x}_{i0}\right\} \geq 1-\alpha, \, i=1,\cdots,m. \quad (3.12)$$

Now suppose $\zeta_r > 0$ is the "external slack" for the rth output. By "external slack" we refer to slack outside the braces. We can choose the value of this external slack so it satisfies

$$\Pr\left\{\sum \tilde{y}_{rj}\lambda_j - \varphi \tilde{y}_{r0} \geq 0\right\} = (1-\alpha) + \zeta_r. \quad (3.13)$$

There must then exist a positive number $s_r^+ > 0$ such that

$$\Pr\left\{\sum \tilde{y}_{rj}\lambda_j - \varphi\tilde{y}_{r0} \geq s_r^+\right\} = 1 + \alpha. \tag{3.14}$$

This positive value of s_r^+ permits a still further increase in $\tilde{y}_{r0}$ for any set of sample observations without worsening any other input or output. It is easy to see that $\zeta_r = 0$ if and only if $s_r^+ = 0$.

In a similar manner, suppose $\xi_i > 0$ represents "external slack" for the ith input chance constraint. We choose its value to satisfy

$$\Pr\left\{\sum_{j=1}^{n} \tilde{x}_{ij}\lambda_j - \tilde{x}_{i0} \leq 0\right\} = (1 - \alpha) + \xi_i. \tag{3.15}$$

There must then exist a positive number $s_i^- > 0$ such that

$$\Pr\left\{\sum_{j=1}^{n} \tilde{x}_{ij}\lambda_j + s_i^- \leq \tilde{x}_{i0}\right\} = 1 - \alpha. \tag{3.16}$$

Such a positive value of s_i^- permits a decrease in $\tilde{x}_{i0}$ for any sample without worsening any other input or output to the indicated probabilities. It is easy to show that $\xi_i = 0$ if and only if $s_i^- = 0$.

We again introduce the non-Archimedean infinitesimal, $\varepsilon > 0$, and extend (3.11) so that stochastic efficiencies and inefficiencies can be characterized by the following model:

$$\begin{cases}
\max \quad \varphi + \varepsilon \left(\sum_{r=1}^{s} s_r^+ + \sum_{i=1}^{m} s_i^-\right) \\
\text{subject to:} \\
\Pr\left\{\sum_{j=1}^{n} \tilde{y}_{rj}\lambda_j - \varphi\tilde{y}_{r0} \geq s_r^+\right\} = 1 - \alpha, r = 1, \cdots, s \\
\Pr\left\{\sum_{j=1}^{n} \tilde{x}_{ij}\lambda_j + s_i^- \leq \tilde{x}_{i0}\right\} = 1 - \alpha, i = 1, \cdots, m \\
\sum_{j=1}^{n} \lambda_j = 1, \\
\lambda_j \geq 0, \ j = 1, \cdots, n \\
s_r^+ \geq 0, \ r = 1, \cdots, s \\
s_i^- \geq 0, \ i = 1, \cdots, m.
\end{cases} \tag{3.17}$$

This leads to the following modification of Definition 3.2.

3.2 Stochastic DEA Models

Definition 3.3. DMU_0 is stochastic efficient if and only if the following two conditions are both satisfied:

1. $\varphi^* = 1$.
2. $s_r^{+*} = 0, s_i^{-*} = 0$ for all $r = 1, 2, \cdots, s$ and $i = 1, 2, \cdots, m$.

This definition aligns more closely with Definition 2.4 since the $\varepsilon > 0$ in the objective of (3.17) makes it unnecessary to refer to all optimal solutions, as in Definition 3.2. It differs from Definition 2.4, however, in that it refers to stochastic characterizations. Thus, even when the conditions of Definition 3.3 are satisfied, there is a chance (determined by the choice of α) that the thus characterized DMU_0 is not efficient.

The stochastic model in (3.17) is evidently a generalization of the BCC model. Assume that inputs and outputs are random variables with a multivariate normal distribution and known parameters. The deterministic equivalent for model (3.17) is as follows:

$$\begin{cases} \max \ \varphi + \varepsilon \left(\sum_{r=1}^{s} s_r^+ + \sum_{i=1}^{m} s_i^- \right) \\ \text{subject to:} \\ \varphi y_{rn} - \sum_{j=1}^{n} y_{rj} \lambda_j + s_r^+ - \Phi^{-1}(\alpha) \sigma_r^0(\varphi, \lambda) = 0, \ r = 1, \cdots, s \\ \sum_{j=1}^{n} x_{ij} \lambda_j + s_i^- - \Phi^{-1}(\alpha) \sigma_i^I(\lambda) = x_{i0}, \ i = 1, \cdots, m \\ \sum_{j=1}^{n} \lambda_j = 1 \\ \lambda_j \geq 0, \ j = 1, \cdots, n \\ s_r^+ \geq 0, \ r = 1, \cdots, s \\ s_i^- \geq 0, \ i = 1, \cdots, m \end{cases} \quad (3.18)$$

where Φ is the standard normal distribution function and Φ^{-1}, its inverse, is the so-called fractile function. Finally,

$$\left(\sigma_r^0(\varphi, \lambda) \right)^2$$
$$= \sum_{i \neq 0} \sum_{j \neq 0} \lambda_i \lambda_j \operatorname{Cov}(\tilde{y}_{ri}, \tilde{y}_{rj}) + 2(\lambda_0 - \varphi) \sum_{i \neq 0} \lambda_i \operatorname{Cov}(\tilde{y}_{ri}, \tilde{y}_{r0})$$
$$+ (\lambda_0 - \varphi)^2 \operatorname{Var}(\tilde{y}_{r0})$$

and

$$\left(\sigma_i^I(\lambda) \right)^2$$
$$= \sum_{j \neq 0} \sum_{k \neq 0} \lambda_i \lambda_k \operatorname{Cov}(\tilde{x}_{ij}, \tilde{x}_{ik}) + 2(\lambda_0 - 1) \sum_{j \neq 0} \lambda_j \operatorname{Cov}(\tilde{x}_{ij}, \tilde{x}_{i0})$$
$$+ (\lambda_0 - 1)^2 \operatorname{Var}(\tilde{x}_{i0})$$

where we have separated out the terms for DMU_0 because they appear on both sides of the expressions in (3.17). Thus, φ^*, s_r^{+*}, and s_i^{-*} can be determined from (3.18) where the data (means and variances) are all assumed to be known.

Let us simplify our assumptions in a manner that will enable us to relate what we are doing to other areas such as the sensitivity analysis research in DEA that is reported in Cooper et al. [15]. Therefore, assume that only DMU_0 has random variations in its inputs and outputs and they are statistically independent. In this case, model (3.18) can be written in the following simpler form:

$$\begin{cases} \max \ \varphi + \varepsilon \left(\sum_{r=1}^{s} s_r^+ + \sum_{i=1}^{m} s_i^- \right) \\ \text{subject to:} \\ \varphi y'_{r0} - \sum_{j=1}^{n} y'_{rj} \lambda_j + s_r^+ = 0, \quad r = 1, \cdots, s \\ \sum_{j=1}^{n} x'_{ij} \lambda_j + s_i^- = x'_{i0}, \quad i = 1, \cdots, m \\ \sum_{j=1}^{n} \lambda_j = 1 \\ \lambda_j \geq 0, \ j = 1, \cdots, n \\ s_r^+ \geq 0, \ r = 1, \cdots, s \\ s_i^- \geq 0, \ i = 1, \cdots, m \end{cases} \quad (3.19)$$

where

$$y'_{r0} = y_{r0} - \sigma_{r0}^o \Phi^{-1}(\alpha), \ r = 1, \cdots, s,$$

$$y'_{rj} = y_{rj}, j \neq 0, \ r = 1, \ldots, s,$$

$$x'_{i0} = x_{i0} + \sigma_{i0}^I \Phi^{-1}(\alpha), \ i = 1, \cdots, m,$$

$$x'_{ij} = x_{ij}, j \neq 0, \ i = 1, \cdots, m,$$

$$\sigma_{r0}^o = \sqrt{\text{Var}(y_{r0})},$$

$$\sigma_{i0}^I = \sqrt{\text{Var}(x_{i0})}.$$

Reasons for us to consider random variations only in DMU_0 are as follows: First, treating more than one DMU in this manner leads to deterministic equivalents with the more complicated relations that have been discussed in detail in Cooper et al. [16–18]. The simpler approach used here allows us to arrive at analytical results and characterizations in a straightforward manner. Second, it opens possible new routes for effecting sensitivity analyses. We are referring to the sensitivity analyses that are to be found in Charnes and Neralic [6], Charnes et al. [9, 10], and Seiford and Zhu [31]. In the terminology of the survey article by Cooper et al. [15], these sensitivity analyses are directed to analyzing allowable limits of data variations for only one DMU at a time and hence contrast with other approaches to sensitivity

3.2 Stochastic DEA Models

analysis in DEA that allow all data for all DMUs to be varied simultaneously until at least one DMU changes its status from efficient to inefficient, or vice versa. These sensitivity analyses are entirely deterministic. Our chance-constrained approach can be implemented by representations that are similar in form to those used in sensitivity analysis, but the conceptual meanings are different. A chance-constrained programming problem can be solved by a deterministic equivalent, as we have just shown, but the issue originally addressed in the chance-constrained formulation is different and this introduces elements, such as the risk associated with α, that are nowhere present in these sensitivity analyses.

Theorem 3.3 (Cooper et al. [16, 17]). *For $\alpha = 0.5$, the inefficiency vs. efficiency classification of DMU_0 in input-output mean model (2.12) is the same as in stochastic model (3.17).*

Theorem 3.4 (Cooper et al. [16, 17]). *For $0 < \alpha < 0.5$.*

(a) *Suppose DMU_0 is efficient with $DMU_0 \in E \bigcup E'$ in input-output mean model (2.12) and then $DMU_0 \in E$ in stochastic model (3.17).*
(b) *Suppose $DMU_0 \in F$ in input-output mean model (2.12) and then $DMU_0 \in E$ in stochastic model (3.17).*
(c) *Suppose $DMU_0 \in N$ in input-output mean model (2.12) and then $DMU_0 \in N$ in stochastic model (3.17) if $\sigma_{i0}^I < \eta_i^{-*}/\left(-\phi^{-1}(\alpha)\right)$ and $\sigma_{r0}^o < \eta_r^{+*}/\left(-\phi^{-1}(\alpha)\right)$, where for $\alpha < 0.5$ we have $\Phi^{-1}(\alpha) < 0$. Here $\sum_{r=1}^{s}\eta_r^{+*} + \sum_{i=1}^{m}\eta_i^{-*}$ is the optimal value of*

$$\begin{cases} \max \sum_{r=1}^{s}\eta_r^+ + \sum_{i=1}^{m}\eta_i^- \\ \text{subject to:} \\ \sum_{j=1}^{n} y_{rj}\lambda_j - \eta_r^+ \geq y_{r0}, r = 1, \cdots, s \\ \sum_{j=1}^{n} x_{ij}\lambda_j + \eta_i^- \leq x_{i0}, i = 1, \cdots, m \\ \sum_{j=1}^{n} \lambda_j = 1, \\ \lambda_j \geq 0, j = 1, \cdots, n \\ \eta_r^+ \geq 0, r = 1, \cdots, s \\ \eta_i^- \geq 0, i = 1, \cdots, m. \end{cases} \quad (3.20)$$

Theorem 3.5 (Cooper et al. [16, 17]). *For $0.5 < \alpha < 1$.*

(a) *Suppose $DMU_0 \in E$ in input-output mean model (2.12) and then $DMU_0 \in E$ in stochastic model (3.17) if*

$$\sum_{r=1}^{s}\sigma_{r0}^o + \sum_{i=1}^{m}\sigma_{i0}^I < \left(\sum_{r=1}^{s}\theta_r^{+*} + \sum_{i=1}^{m}\theta_i^{-*}\right)/\Phi^{-1}(\alpha)$$

where $\sum_{r=1}^{s} \theta_r^{+*} + \sum_{i=1}^{m} \theta_i^{-*}$ is the optimal value of

$$\begin{cases} \min \sum_{r=1}^{s} \theta_r^+ + \sum_{i=1}^{m} \theta_i^- \\ \text{subject to:} \\ \quad \sum_{\substack{j=0 \\ j \neq 0}}^{n} y_{rj} \lambda_j \geq y_{r0} - \theta_r^+, r = 1, \cdots, s \\ \quad \sum_{\substack{j=0 \\ j \neq 0}}^{n} x_{ij} \lambda_j \leq x_{i0} + \theta_i^-, i = 1, \cdots, m \\ \quad \sum_{j=1}^{n} \lambda_j = 1 \\ \quad \lambda_j \geq 0, \ j = 1, \cdots, n \\ \quad \eta_r^+ \geq 0, \ r = 1, \cdots, s \\ \quad \eta_i^- \geq 0, \ i = 1, \cdots, m. \end{cases} \quad (3.21)$$

(b) Suppose $DMU_0 \in E' \bigcup F \bigcup N$ in input-output mean model (2.12) and then $DMU_0 \in N$ in stochastic model (3.17).

3.2.2 Satisfying DEA Models

In the previous sections, we have discussed joint chance-constrained DEA formulations and "E-model" chance-constrained DEA forms. In this section, we would like to discuss another type of chance-constrained DEA model, referred to as "P-models" in the chance-constrained programming literature. We can also refer to them as "satisfying" DEA models, as drawn from Cooper et al. [13].

We start by introducing the following version of a P-Model which we use to adapt the usual definitions of "DEA efficiency" to a chance-constrained programming context:

$$\begin{cases} \max \ \Pr \left\{ \dfrac{\sum_{r=1}^{s} u_r \tilde{y}_{ro}}{\sum_{i=1}^{m} v_i \tilde{x}_{io}} \geq 1 \right\} \\ \text{subject to:} \\ \quad \Pr \left\{ \dfrac{\sum_{r=1}^{s} u_r \tilde{y}_{rj}}{\sum_{i=1}^{m} v_i \tilde{x}_{ij}} \leq 1 \right\} \geq 1 - \alpha_j, \ j = 1, \cdots, n \\ \quad u_r \geq 0, \ r = 1, \cdots, s \\ \quad v_i \geq 0, \ i = 1, \cdots, m \end{cases} \quad (3.22)$$

in which Pr means probability measure and u_r and $v_i \geq 0$ are the virtual multipliers to be determined by solving the above problem. This model evidently builds upon

3.2 Stochastic DEA Models

the CCR model of DEA, as derived in Charnes et al. [7], with the ratio in the objective representing output and input values for DMU_0, which is also included in the jth DMUs with output-to-input ratios represented as chance constraints, $j = 1, \cdots, n$.

Evidently, the constraints in (3.22) are satisfied by choosing $u_r = 0$ and $v_i > 0$ for all r and i. Hence, for continuous distributions like the ones considered in this section, it is not vacuous to write

$$\Pr\left\{\frac{\sum_{r=1}^{s} u_r^* \tilde{y}_{ro}}{\sum_{i=1}^{m} v_i^* \tilde{x}_{io}} \leq 1\right\} + \Pr\left\{\frac{\sum_{r=1}^{s} u_r^* \tilde{y}_{ro}}{\sum_{i=1}^{m} v_i^* \tilde{x}_{io}} \geq 1\right\} = 1$$

or

$$\Pr\left\{\frac{\sum_{r=1}^{s} u_r^* \tilde{y}_{ro}}{\sum_{i=1}^{m} v_i^* \tilde{x}_{io}} \leq 1\right\} = 1 - \alpha^* \geq 1 - \alpha_0.$$

Here, $*$ refers to an optimal value so α^* is the probability of achieving a value of at least unity with this choice of weights and $1 - \alpha^*$ is, therefore, the probability of failing to achieve this value.

To see how these formulations may be used, we note that we must have $\alpha_0 \geq \alpha^*$ since $1 - \alpha^*$ is prescribed in the constraint for $j = 0$ as the chance allowed for characterizing the $\tilde{y}_{ro}$, $\tilde{x}_{io}$ values as inefficient. More formally, we introduce the following stochasticized definition of efficiency.

Definition 3.4. DMU_0 is "stochastic efficient" if and only if $\alpha^* = \alpha_0$.

This opens a variety of new directions for research and potential uses of DEA. Before indicating some of these possibilities, however, we replace (3.22) with the following:

$$\begin{cases} \max \Pr\left\{\dfrac{\sum_{r=1}^{s} u_r \tilde{y}_{ro}}{\sum_{i=1}^{m} v_i \tilde{x}_{io}} \geq 1\right\} \\ \text{subject to:} \\ \Pr\left\{\dfrac{\sum_{r=1}^{s} u_r \tilde{y}_{rj}}{\sum_{i=1}^{m} v_i \tilde{x}_{ij}} \leq 1\right\} + \Pr\left\{\dfrac{\sum_{r=1}^{s} u_r \tilde{y}_{ro}}{\sum_{i=i}^{m} v_i \tilde{x}_{io}} \geq 1\right\} \geq 1, \; j = 1, \cdots, n \\ u_r \geq 0, \; r = 1, \cdots, s \\ v_i \geq 0, \; i = 1, \cdots, m. \end{cases} \quad (3.23)$$

This simpler model makes it easier to see what is involved in uses of these DEA formulations. It also enables us to examine potential uses in a simplified manner.

First, as is customary in CCP, it is assumed that the behavior of the random variables is governed by a known multivariate distribution. Hence, we can examine the value of α^* even before the data are generated. If this value is too small, then one can signal central management, say, that the situation for DMU_0 needs to be examined in advance because there is a probability of at least $1 - \alpha^* \geq 1 - \alpha_0$ that it will not perform efficiently.

If we define a "rule" as a "chance constraint which is to hold with probability one," then we can regard a "policy" as "a chance constraint which is to hold with probability $0.5 < \alpha^* < 1$." Implementation of a "policy" allows for deviations which can require managerial attention, whereas a "rule" may be administered in clerical fashion since no exceptions are to be permitted. Notice, too, that a policy may be identified and evaluated by reference to ex post data, as in an accounting or performance audit, in order to see whether the corresponding actions had been taken sufficiently frequently or whether some "policy" other than the intended one had prevailed; see Cooper and Ijiri [11].

The following model represents an evident generalization of (3.22):

$$\begin{cases} \max \ \Pr\left\{ \dfrac{\sum_{r=1}^{s} u_r \tilde{y}_{ro}}{\sum_{i=1}^{m} v_i \tilde{x}_{io}} \geq \eta_o \right\} \\ \text{subject to:} \\ \quad \Pr\left\{ \dfrac{\sum_{r=1}^{s} u_r \tilde{y}_{rj}}{\sum_{i=1}^{m} v_i \tilde{x}_{ij}} \leq \eta_j \right\} \geq 1 - \alpha_j, \ j = 1, \cdots, n \\ \quad \Pr\left\{ \dfrac{\sum_{r=1}^{s} u_r \tilde{y}_{rj}}{\sum_{i=1}^{m} v_i \tilde{x}_{ij}} \geq \eta_j \right\} \geq 1 - \alpha_j, \ j = n+1, \cdots, n+k \\ \quad u_r \geq 0, \ r = 1, \cdots, s \\ \quad v_i \geq 0, \ i = 1, \cdots, m. \end{cases} \quad (3.24)$$

In a similar manner to the analysis of model (3.22), it is obvious that the first $j = 1, \cdots, n$ constraints in (3.24) are satisfied by choosing $u_r = 0$, and $v_i > 0$ for all r and i. For an optimal solution (u^*, v^*), we must have

$$\Pr\left\{ \dfrac{\sum_{r=1}^{s} u_r^* \tilde{y}_{ro}}{\sum_{i=1}^{m} v_i^* \tilde{x}_{io}} \leq \eta_o \right\} = 1 - \alpha^* \geq 1 - \alpha_0.$$

Therefore, we introduce the following stochasticized definitions of efficiency.

Definition 3.5 (Stochastic Efficiency). If $\eta_{jo} = \eta_o = 1$, DMU_0 is "stochastically efficient" if and only if $\alpha^* = \alpha_0$.

3.2 Stochastic DEA Models

Definition 3.6 (Satisficing Efficiency). If $\eta_{j0} = \eta_0 < 1$, DMU$_0$ is "satisficing efficient" if and only if $\alpha^* = \alpha_0$.

Now we assume $\alpha_j < 0.5$. Utilizing techniques in chance-constrained programming theory, a deterministic equivalent of (3.24) is then as follows:

$$\begin{cases} \max \gamma \\ \text{subject to:} \\ \quad u^T y_0 - \eta_0 v^T x_0 \geq \Phi^{-1}(\gamma) \sqrt{u^T \Sigma_0 u} \\ \quad u^T y_j - \eta_j v^T x_j - \Phi^{-1}(\alpha_j) \eta_j \leq 0, \ j = 1, \cdots, n \\ \quad \eta_j^2 - u^T \Sigma_j u \geq 0, \ j = 1, \cdots, n \\ \quad u \geq 0 \\ \quad v \geq 0 \\ \quad \eta \geq 0 \end{cases} \quad (3.25)$$

where $\Sigma_j = \left(\text{Cov}\left(\tilde{y}_{ij}, \tilde{y}_{kj} \right) \right)$.

This is a nonlinear and nonconvex programming problem. However, let us consider the following quadratic programming problem:

$$\begin{cases} \max \delta \\ \text{subject to:} \\ \quad \mu^T y_0 - \eta_0 v^T x_0 \geq \delta \\ \quad \mu^T \Sigma_0 \mu \geq 1 \\ \quad \mu^T y_j - \eta_j v^T x_j - \Phi^{-1}(\alpha_j) \xi_j \leq 0, \ j = 1, \cdots, n \\ \quad \xi_j^2 - \mu^T \Sigma_j \mu \geq 0, \ j = 1, \cdots, n \\ \quad \mu \geq 0 \\ \quad v \geq 0 \\ \quad \xi \geq 0. \end{cases} \quad (3.26)$$

It is easy to show that if δ^* is the optimal value of (3.26) and γ^* is the optimal value of (3.25), then we have $\Phi(\delta^*) = \gamma^*$. Therefore, we have the following result.

Theorem 3.6 (Cooper et al. [13]). *If $\eta_{j0} = \eta_0 = 1$ ($\eta_{j0} = \eta_0 < 1$), DMU$_0$ is stochastically (satisficing) efficient if and only if $\Phi(\delta^*) = \alpha_0$.*

Now let us discuss the case of $\Phi(\delta^*) < \alpha_0$. In this case, the risk of failing to satisfy the constraints for DMU$_{j_0}$ falls below the level which was specified as satisfactory. To state this in a more positive manner, let us consider the fact that

$$\Pr\left\{ \frac{\sum_{r=1}^{s} u_r^* \tilde{y}_{r0}}{\sum_{i=1}^{m} v_i^* \tilde{x}_{i0}} \leq \eta_0 \right\} + \Pr\left\{ \frac{\sum_{r=1}^{s} u_r^* \tilde{y}_{r0}}{\sum_{i=i}^{m} v_i^* \tilde{x}_{i0}} \geq \eta_0 \right\} = 1.$$

Therefore,

$$\Pr\left\{\frac{\sum_{r=1}^{s} u_r^* \tilde{y}_{r0}}{\sum_{i=1}^{m} v_i^* \tilde{x}_{i0}} \leq \eta_0\right\} = 1 - \Phi(\delta^*) > 1 - \alpha_0,$$

which leads to the following corollary to the above theorem:

Corollary 3.1. *If $\eta_{j0} = \eta_0 = 1$ ($\eta_{j0} = \eta_0 < 1$), DMU_0 is stochastically (satisficing) inefficient if and only if $\Phi(\delta^*) < \alpha_0$.*

Note: (1) If $\alpha_j = 0.5$ for DMU_j, the constraint $\xi_j^2 - \mu^T \sum_j \mu \geq 0$ should be deleted from model (3.26); (2) if $\alpha_j > 0.5$ for DMU_j, the constraint $\xi_j^2 - \mu^T \sum_j \mu \geq 0$ should be changed to be $\xi_j^2 - \mu^T \sum_j \mu \leq 0$.

3.3 Stochastic DEA Ranking Criteria

This section will introduce some fully ranking methods in stochastic DEA. Four types of stochastic DEA fully ranking criteria are to be investigated.

3.3.1 Expected Ranking Criterion

The essential idea of the stochastic expected DEA model is to optimize the expected value of $\dfrac{v^T \tilde{y}_0}{u^T \tilde{x}_0}$ subject to some chance constraints, then we have the first type of the stochastic DEA model:

$$\begin{cases} \theta = \max_{u,v} E\left[\dfrac{v^T \tilde{y}_0}{u^T \tilde{x}_0}\right] \\ \text{subject to:} \\ \quad \Pr\{v^T \tilde{y}_k \leq u^T \tilde{x}_k\} \geq \alpha, \ k = 1, 2, \cdots, n \\ \quad u \geq 0 \\ \quad v \geq 0 \end{cases} \quad (3.27)$$

in which $\alpha \in (0.5, 1]$.

Definition 3.7. *A vector $(u, v) \geq 0$ is called a feasible solution to the stochastic programming model (3.27) if*

$$\Pr\{v^T \tilde{y}_k \leq u^T \tilde{x}_k\} \geq \alpha \quad (3.28)$$

for $k = 1, 2, \cdots, n$.

3.3 Stochastic DEA Ranking Criteria

Definition 3.8. A feasible solution (u^*, v^*) is called an expected optimal solution to the stochastic programming model (3.27) if

$$E\left[\frac{v^{*T}\tilde{y}_0}{u^{*T}\tilde{x}_0}\right] \geq E\left[\frac{v^T\tilde{y}_0}{u^T\tilde{x}_0}\right] \quad (3.29)$$

for any feasible solution (u, v).

Expected Ranking Criterion: The greater the optimal objective value is, the more efficient DMU_0 is ranked.

3.3.2 Optimistic Ranking Criterion

Chance-constrained programming (CCP), which was initialized by Charnes and Cooper [5], offers a powerful means for modeling stochastic decision systems. The essential idea of chance-constrained programming is to optimize some critical value with a given confidence level subject to some chance constraints. Assuming that the decision makers want to maximize the optimistic value of the stochastic objective at given confidence level, we have the second type of stochastic DEA model:

$$\begin{cases} \max \overline{f} \\ \text{subject to:} \\ \quad \Pr\left\{\frac{v^T\tilde{y}_0}{u^T\tilde{x}_0} \leq \overline{f}\right\} \leq \alpha \\ \quad \Pr\left\{v^T\tilde{y}_j \leq u^T\tilde{x}_j\right\} \geq \alpha, \ j = 1, 2, \cdots, n \\ \quad u \geq 0 \\ \quad v \geq 0 \end{cases} \quad (3.30)$$

in which $\alpha \in [0.5, 1]$.

Definition 3.9. Let ξ and η be random variables. We say $\xi > \eta$ if and only if, for some predetermined confidence level $\alpha \in (0, 1]$, we have

$$\sup\{r \mid \Pr\{\xi \leq r\} \leq \alpha\} > \sup\{r \mid \Pr\{\eta \leq r\} \leq \alpha\}. \quad (3.31)$$

Ranking Criterion: The greater the optimal objective is, the more efficient DMU_0 is ranked.

3.3.3 Maximal Chance Ranking Criterion

Sometimes the decision maker may want to maximize the chance of satisfying the event $\frac{v^T \tilde{y}_0}{u^T \tilde{x}_0} \geq 1$. In order to model this type of decision system, Liu [26–28] provided the dependent-chance programming (DCP). Here we carried out the DCP model into the stochastic DEA as follows:

$$\begin{cases} \max\limits_{u,v} \theta = \Pr\left\{ \dfrac{v^T \tilde{y}_0}{u^T \tilde{x}_0} \geq 1 \right\} \\ \text{subject to:} \\ \quad \Pr\left\{ v^T \tilde{y}_k \leq u^T \tilde{x}_k \right\} \geq 1 - \alpha, \ k = 1, 2, \cdots, n \\ \quad u \geq 0 \\ \quad v \geq 0 \end{cases} \quad (3.32)$$

in which $\alpha \in (0, 0.5]$.

Definition 3.10. A feasible solution (u^*, v^*) is called a maximal chance optimal solution to the stochastic programming model (3.32) if

$$\Pr\left\{ \frac{v^{*T} \tilde{y}_0}{u^{*T} \tilde{x}_0} \geq 1 \right\} \geq \Pr\left\{ \frac{v^T \tilde{y}_0}{u^T \tilde{x}_0} \geq 1 \right\} \quad (3.33)$$

for any feasible solution (u, v).

Ranking Method: The greater the optimal objective is, the more efficient DMU$_0$ is ranked.

3.3.4 Hurwicz Ranking Criterion

Most of the DEA models evaluate the distance of DMU$_0$ to an efficient frontier, which can be considered an optimistic method:

$$\begin{cases} \theta_1 = \max \ \sum\limits_{i=1}^{p} s_i^- + \sum\limits_{j=1}^{q} s_j^+ \\ \text{subject to:} \\ \quad \Pr\left\{ \sum\limits_{k=1}^{n} \tilde{x}_{ki} \lambda_k \leq \tilde{x}_{0i} - s_i^- \right\} \geq \alpha, \ i = 1, 2, \cdots, p \\ \quad \Pr\left\{ \sum\limits_{k=1}^{n} \tilde{y}_{kj} \lambda_k \geq \tilde{y}_{0j} + s_j^+ \right\} \geq \alpha, \ j = 1, 2 \cdots, q \\ \quad \sum\limits_{k=1}^{n} \lambda_k = 1 \\ \quad \lambda_k \geq 0, \quad k = 1, 2, \cdots, n \\ \quad s_i^- \geq 0, \quad i = 1, 2 \cdots, p \\ \quad s_j^+ \geq 0, \quad j = 1, 2, \cdots, q. \end{cases} \quad (3.34)$$

3.3 Stochastic DEA Ranking Criteria

Definition 3.11. DMU_0 is α-efficient if s_i^{-*} and s_j^{+*} are zero for $i = 1, 2, \cdots, p$ and $j = 1, 2 \cdots, q$, where s_i^{-*} and s_j^{+*} are optimal solutions of (3.34).

At the other hand, Jahanshahloo and Afzalinejad [24] proposed a ranking method which uses the distance to the inefficient frontier as its efficiency value. Similar to model (3.34), the pessimistic model, which considers the total distances to an inefficient frontier, can be given as

$$\begin{cases} \theta_2 = \max \sum_{i=1}^{p} s_i^- + \sum_{j=1}^{q} s_j^+ \\ \text{subject to:} \\ \quad \Pr\left\{\sum_{k=1}^{n} \tilde{x}_{ki}\lambda_k \geq \tilde{x}_{0i} + s_i^-\right\} \geq \alpha, \quad i = 1, 2, \cdots, p \\ \quad \Pr\left\{\sum_{k=1}^{n} \tilde{y}_{kj}\lambda_k \leq \tilde{y}_{0j} - s_j^+\right\} \geq \alpha, \quad j = 1, 2\cdots, q \\ \quad \sum_{j=1}^{n} \lambda_j = 1 \\ \quad \lambda_j \geq 0, \quad j = 1, 2, \cdots, n \\ \quad s_i^- \geq 0, \quad i = 1, 2, \cdots, p \\ \quad s_j^+ \geq 0, \quad j = 1, 2\cdots, q. \end{cases} \quad (3.35)$$

As a result, this section wants to rank all the DMUs by the Hurwicz criterion [22, 23], which attempts to find a middle ground between the optimistic and pessimistic criteria.

Definition 3.12 (α-Inefficiency). DMU_0 is α-inefficient if s_i^{-*} and s_j^{+*} are zero for $i = 1, 2, \cdots, p$ and $j = 1, 2 \cdots, q$, where s_i^{-*} and s_j^{+*} are optimal solutions of (3.35).

Since $j = 0$ is one of the DMU_j, we can always get a solution with $\lambda_0 = 1, \lambda_j = 0$ ($j \neq 0$) and all slacks zero. Thus fuzzy DEA models (3.34) and (3.35) have feasible solution and the optimal value $s_i^{-*} = s_j^{+*} = 0$ for all i, j.

Above mentioned two models are both extreme cases: One is too optimistic and the other is too pessimistic. Thus, we employ the Hurwicz criterion, suggested by Leonid Hurwicz [22, 23] in 1951, which incorporates a measure of both by assigning a certain percentage weight λ to θ_1^* and $1 - \lambda$ to $-\theta_2^*$, $0 \leq \lambda \leq 1$:

$$\theta^* = \lambda \theta_1^* + (1-\lambda)(-\theta_2^*) \quad (3.36)$$

which can be rewritten as

$$\theta^* = \lambda \theta_1^* - (1-\lambda)\theta_2^*. \quad (3.37)$$

Ranking Criterion: The greater the value θ^* is, the less efficient DMU_0 is ranked.

In the Hurwicz criterion, the parameter $\lambda \in [0, 1]$, which reflects the degree of the decision maker's optimism, must be determined by the decision maker. Generally speaking, it is difficult to determine the appropriate λ for decision makers, since it varies from person to person. By varying the parameter λ, the Hurwicz criterion becomes various models, e.g., when $\lambda = 1$, the criterion is the traditional DEA model (3.34); when $\lambda = 0$, it degenerates to model (3.35). This fact means that the Hurwicz criterion is fairly flexible.

In some cases, $\theta_1^* = 0$ and $\theta_2^* = 0$, and then the ranking value $\theta^* = 0$. It says DMU_0 is both efficient and inefficient. This happens when DMU_0 is the best in some inputs and/or outputs while it is the worst in some other inputs and/or outputs. For example, if DMU_0 is the only DMU that has the largest value for input 1 and has the least amount of input 2, DMU_0 is both efficient and inefficient.

References

1. Aigner D, Lovell CAK, Schmidt P (1977) Formulation and estimation of stochastic frontier production function models. J Econ 6:21–37
2. Banker RD (1986) Stochastic data envelopment analysis. Carnegie-Mellon University, Pittsburgh
3. Banker RD (1993) Maximum likelihood, consistency and DEA: statistical foundations. Manag Sci 39:1265–1273
4. Banker RD, Charnes A, Cooper WW (1984) Some models for estimating technical and scale efficiencies in data envelopment analysis. Manag Sci 30:1078–1092
5. Charnes A, Cooper W (1961) Management models and industrial applications of linear programming. Wiley, New York
6. Charnes A, Neralic L (1990) Sensitivity analysis of the additive model in data envelopment analysis. Eur J Oper Res 48:332–341
7. Charnes A, Cooper WW, Rhodes E (1978) Measuring the efficiency of decision making units. Eur J Oper Res 2:429–444
8. Charnes A, Cooper WW, Golany B, Seiford L, Stutz J (1985) Foundations of data envelopment analysis for Pareto-Koopmans efficient empirical production functions. J Econom 30:91–107
9. Charnes A, Haag S, Jaska P, Semple J (1992) Sensitivity of efficiency classifications in the additive model of data envelopment analysis. Int J Syst Sci 23:789–798
10. Charnes A, Rousseau J, Semple J (1996) Sensitivity and stability of efficiency classifications in data envelopment analysis. J Product Anal 7:5–18
11. Cooper WW, Ijiri Y (eds) (1983) Kohler's dictionary for accountants, 6th edn. Prentice Hall, Englewood Cliffs
12. Cooper WW, Huang ZM, Li SX (1996) Satisfying DEA models under chance constraints. Ann Oper Res 66:279–95
13. Cooper WW, Thompson RG, Thrall RM (1996) Introduction: extensions and new developments in DEA. Ann Oper Res 66:3–46
14. Cooper WW, Huang ZM, Lelas V, Li SX, Olesen OB (1998) Chance constrained programming formulations for stochastic characterizations of efficiency and dominance in DEA. J Product Anal 9:53–79
15. Cooper WW, Li S, Seiford LM, Tone K, Thrall RM, Zhu J (2001) Sensitivity and stability analysis in DEA: some recent developments. J Product Anal 15:217–246

References

16. Cooper WW, Deng H, Huang ZM, Li SX (2002) Chance constrained programming approaches to technical efficiencies and inefficiencies in stochastic data envelopment analysis. J Oper Res Soc 53:1347–1356
17. Cooper WW, Lelas V, DSullivan DM (2002) Chance-constrained programming and skewed distributions of matrix coefficients and applications to environmental regulatory activities. In: Cawrence KD, Kleinbore RK (eds) Science: in productivity, finance and management. Elsevier, Oxford
18. Cooper WW, Deng H, Huang ZM, Li SX (2003) Chance constrained programming approaches to congestion in stochastic data envelopment analysis. Eur J Oper Res 155:231–238
19. Gong BH, Sickles RC (1992) Finite sample evidence on the performance of stochastic frontiers and data envelopment analysis using pane: data. J Econom 51:27–56
20. Grosskopf S (1996) Statistical inference and nonparametric efficiency: a selective survey. J Product Anal 7:161–176
21. Horace WC, Schmidt P (1996) Confidence statements for efficiency estimates from stochastic frontier models. J Product Anal 7:257–282
22. Hurwicz L (1951) Optimality criteria for decision making under ignorance. Cowles Commission Discussion Paper, Chicago, 370
23. Hurwicz L (1951) Some specification problems and application to econometric models (abstract). Econometrica 19:343–344
24. Jahanshahloo GR, Afzalinejad M (2006) A ranking method based on a full-inefficient frontier. Appl Math Model 30:248–260
25. Land KC, Lovell CAK, Thore S (1993) Chance constrained data envelopment analysis. Manag Decis Econ 14:541–554
26. Liu B (1997) Dependent-chance programming: a class of stochastic programming. Comput Math Appl 34(12):89–104
27. Liu B (1999) Dependent-chance programming with fuzzy decisions. IEEE Trans Fuzzy Syst 7:354–360
28. Liu B (2002) Random fuzzy dependent-chance programming and its hybrid intelligent algorithm. Inf Sci 141(3–4):259–271
29. Olesen O (2006) Comparing and combining two approaches for chance constrained DEA. J Product Anal 26:103–119
30. Olesen O, Petersen NC (1995) Chance constrained efficiency evaluation. Manag Sci 141:442–457
31. Seiford LM, Zhu J (1998) Stability regions for maintaining efficiency in data envelopment analysis. Eur J Oper Res 108(1):127–139
32. Sengupta JK (1982) Efficiency measurement in stochastic input-output systems. Int J Syst Sci 13:273–287
33. Simar L (1996) Aspects of statistical analysis in DEA-type frontier models. J Product Anal 7:175–185
34. Simar L, Wilson PW (1988) Sensitivity analysis of efficiency scores: how to bootstrap in nonparametric frontier models. Manag Sci 44:49–61

Chapter 4
Fuzzy DEA

In more general cases, the evaluation data are often collected through investigation into the quality of the natural language such as "good," "medium," or "bad" rather than a specific case. That is, the data are fuzzy. We can find several fuzzy approaches to the efficiency assessment in the DEA literature. Cooper et al. [14,15], one of the DEA initiators, introduced how to deal with imprecise data such as bounded data, ordinal data, and ratio bounded data in DEA. Kao and Liu [29] developed a method to find the membership functions of the fuzzy efficiency scores when some observations are fuzzy numbers. Entani et al. [17] proposed a DEA model with an interval efficiency which is obtained from the pessimistic and the optimistic viewpoints. Since possibility measure (Zadeh [46]) has been widely used in dealing with fuzzy data, many researchers have introduced it into DEA (Guo and Tanaka [24] and Lertworasirikul et al. [32]). However, the possibility measure has no self-duality property, which is needed both in theory and in practice. In order to define a self-dual measure, Liu and Liu [38] presented the concept of credibility measure in 2002. This chapter will give a brief introduction to fuzzy DEA based on credibility measure.

4.1 Symbols and Notations

This section will introduce some symbols and notations which will be used in the following sections:

DMU_i: the ith DMU, $i = 1, 2, \cdots, n$
DMU_0: the target DMU
$\tilde{x}_i = (\tilde{x}_{i1}, \tilde{x}_{i2}, \cdots, \tilde{x}_{ip})$: the fuzzy input vectors of DMU_i, $i = 1, 2, \cdots, n$, respectively
$\tilde{y}_i = (\tilde{y}_{i1}, \tilde{y}_{i2}, \cdots, \tilde{y}_{iq})$: the fuzzy output vectors of DMU_i, $i = 1, 2, \cdots, n$, respectively
$x_0 = (x_{01}, x_{02}, \cdots, x_{0p})$: the input vector of the target DMU_0

$y_0 = (y_{01}, y_{02}, \cdots, y_{0q})$: the output vector of the target DMU_0
$u \in R^{p \times 1}$: the vector of input weights
$v \in R^{q \times 1}$: the vector of output weights

4.2 Fuzzy DEA Models

Since the inputs and outputs are fuzzy variables, the constraints $\sum_{k=1}^{n} \tilde{x}_{ki} \lambda_k \leq \tilde{x}_{0i} - s_i^-$, $i = 1, 2 \cdots, p$ and $\sum_{k=1}^{n} \tilde{y}_{kj} \lambda_k \geq \tilde{y}_{0j} + s_j^+$, $j = 1, 2, \cdots, q$ do not define a deterministic feasible set. A natural idea is to provide a confidence level α, at which it is desired that the fuzzy constraints hold. In other words, the event may not happen within $1 - \alpha$ credibility level. Thus, we have some chance constraints as follows:

$$\text{Cr} \left\{ \sum_{k=1}^{n} \tilde{x}_{ki} \lambda_k \leq \tilde{x}_{0i} - s_i^- \right\} \geq \alpha, \quad i = 1, 2, \cdots, p,$$

$$\text{Cr} \left\{ \sum_{k=1}^{n} \tilde{y}_{kj} \lambda_k \geq \tilde{y}_{0j} + s_j^+ \right\} \geq \alpha, \quad j = 1, 2 \cdots, q \quad (4.1)$$

in which Cr is the credibility measure introduced in Sect. 1.2.

Similar to the deterministic case, Wen et al. [44] proposed the fuzzy DEA model in which the objective is to maximize the total slacks in inputs and outputs subject to the constraints (4.1):

$$\begin{cases} \max \quad \sum_{i=1}^{p} s_i^- + \sum_{j=1}^{q} s_j^+ \\ \text{subject to:} \\ \quad \text{Cr} \left\{ \sum_{k=1}^{n} \tilde{x}_{ki} \lambda_k \leq \tilde{x}_{0i} - s_i^- \right\} \geq \alpha, \quad i = 1, 2, \cdots, p \\ \quad \text{Cr} \left\{ \sum_{k=1}^{n} \tilde{y}_{kj} \lambda_k \geq \tilde{y}_{0j} + s_j^+ \right\} \geq \alpha, \quad j = 1, 2 \cdots, q \\ \quad \sum_{k=1}^{n} \lambda_k = 1 \\ \quad \lambda_k \geq 0, \quad k = 1, 2, \cdots, n \\ \quad s_i^- \geq 0, \quad i = 1, 2 \cdots, p \\ \quad s_j^+ \geq 0, \quad j = 1, 2, \cdots, q. \end{cases} \quad (4.2)$$

Definition 4.1 (Wen et al. [44]). DMU_0 is α-efficient if s_i^{-*} and s_j^{+*} are zero for $i = 1, 2, \cdots, p$ and $j = 1, 2 \cdots, q$, where s_i^{-*} and s_j^{+*} are optimal solutions of (4.2).

4.2 Fuzzy DEA Models

This definition aligns more closely with Definition 2.5. However, it differs in that credibility measure is involved. For instance, as determined by the choice of α, there is a risk that DMU_0 will not be efficient even when the condition of Definition 4.1 is satisfied.

Since $j = 0$ is one of the DMU_j, we can always get a solution with $\lambda_0 = 1, \lambda_j = 0$ ($j \neq 0$) and all slacks zero. Thus, this fuzzy DEA model has feasible solution and the optimal value $s_i^{-*} = s_j^{+*} = 0$ for all i, j.

Definition 4.2 (Liu [35]). Let ξ be a fuzzy variable, and $\alpha \in (0, 1]$. Then

$$\xi_{\sup}(\alpha) = \sup\{r \mid \text{Cr}\{\xi \geq r\} \geq \alpha\}$$

is called the α-optimistic value to ξ, and

$$\xi_{\inf}(\alpha) = \inf\{r \mid \text{Cr}\{\xi \leq r\} \geq \alpha\}$$

is called the α-pessimistic value to ξ.

Theorem 4.1 (Liu [35]). *Let $\xi_{\inf}(\alpha)$ and $\xi_{\sup}(\alpha)$ be the α-pessimistic and α-optimistic values of the fuzzy variable ξ, respectively. Then we have:*

(a) $\xi_{\inf}(\alpha)$ is an increasing and left-continuous function of α.
(b) $\xi_{\sup}(\alpha)$ is a decreasing and left-continuous function of α.
(c) If $c \geq 0$, then $(c\xi)_{\sup}(\alpha) = c\xi_{\sup}(\alpha)$ and $(c\xi)_{\inf}(\alpha) = c\xi_{\inf}(\alpha)$.
(d) If $c < 0$, then $(c\xi)_{\sup}(\alpha) = c\xi_{\inf}(\alpha)$ and $(c\xi)_{\inf}(\alpha) = c\xi_{\sup}(\alpha)$.
(e) $(\xi + \eta)_{\sup}(\alpha) = \xi_{\sup}(\alpha) + \eta_{\sup}(\alpha)$, $(\xi + \eta)_{\inf}(\alpha) = \xi_{\inf}(\alpha) + \eta_{\inf}(\alpha)$.

Theorem 4.2. *The objective value in (4.2) is a decreasing function of α.*

Proof. From the property of $\xi_{\sup}^{\alpha}$, the result can be easily proved.

From Theorem 4.2, we get that the efficiency of DMU_0 is affected by the confidence level α. The bigger the value α is, the more efficient the DMU_0 is.

From the properties of the α-optimistic and α-pessimistic values, we can rewrite the fuzzy DEA model as follows:

$$\begin{cases} \max \quad \sum_{i=1}^{p} s_i^- + \sum_{j=1}^{q} s_j^+ \\ \text{subject to:} \\ \sum_{k=1}^{n} \lambda_k (\tilde{x}_{ki})_{\inf}^{\alpha} + \lambda_0 \left[(\tilde{x}_{0i})_{\sup}^{\alpha} - (\tilde{x}_{0i})_{\inf}^{\alpha} \right] \leq (\tilde{x}_{0i})_{\sup}^{\alpha} - s_i^-, \quad i = 1, 2 \cdots, p \\ \sum_{k=1}^{n} \lambda_k (\tilde{y}_{kj})_{\sup}^{\alpha} + \lambda_0 \left[(\tilde{y}_{0j})_{\inf}^{\alpha} - (\tilde{y}_{0j})_{\sup}^{\alpha} \right] \geq (\tilde{y}_{0j})_{\inf}^{\alpha} + s_j^+, \quad j = 1, 2, \cdots, q \\ \sum_{k=1}^{n} \lambda_k = 1 \\ \lambda_k \geq 0, \quad k = 1, 2, \cdots, n \\ s_i^- \geq 0, \quad i = 1, 2 \cdots, p \\ s_j^+ \geq 0, \quad j = 1, 2, \cdots, q \end{cases} \quad (4.3)$$

which is a linear programming. Thus, it can be easily solved by many traditional methods.

When the inputs and outputs are trapezoidal fuzzy variables, following Liu's book [37], the fuzzy model becomes the following linear programming:

$$\begin{cases} \max \quad \sum_{i=1}^{p} s_i^- + \sum_{j=1}^{q} s_j^+ \\ \text{subject to:} \\ \quad \sum_{k=1}^{n} \lambda_k \left[2(1-\alpha) x_{ki}^c + (2\alpha - 1) x_{ki}^d \right] \\ \quad + \lambda_0 \left[2(1-\alpha) \left(x_{0i}^b - x_{0i}^c \right) + (2\alpha - 1) \left(x_{0i}^a - x_{0i}^d \right) \right] \\ \quad \leq 2(1-\alpha) x_{0i}^b + (2\alpha - 1) x_{0i}^a - s_i^-, i = 1, 2, \cdots, p \\ \quad \sum_{k=1}^{n} \lambda_k \left[2(1-\alpha) y_{kj}^b + (2\alpha - 1) y_{kj}^a \right] \\ \quad + \lambda_0 \left[2(1-\alpha) \left(y_{0j}^c - y_{0j}^b \right) + (2\alpha - 1) \left(y_{0j}^d - y_{0j}^a \right) \right] \\ \quad \geq 2(1-\alpha) y_{0j}^c + (2\alpha - 1) y_{0j}^d + s_j^+, j = 1, 2, \cdots, q \\ \quad \sum_{j=1}^{n} \lambda_k = 1 \\ \quad \lambda_k \geq 0, \quad k = 1, 2, \cdots, n \\ \quad s_i^- \geq 0, \quad i = 1, 2, \cdots, p \\ \quad s_j^+ \geq 0, \quad j = 1, 2 \cdots, q \end{cases} \quad (4.4)$$

in which the trapezoidal fuzzy variable $\left(x_{ki}^a, x_{ki}^b, x_{ki}^c, x_{ki}^d \right)$ represents the ith input of DMU$_k$ and $\left(y_{kj}^a, y_{kj}^b, y_{kj}^c, y_{kj}^d \right)$ represents the jth output of DMU$_k$, respectively.

Example 4.1. This example wants to give some comparisons when using different measures to the fuzzy DEA model (4.2). Let us consider two DMUs with one input x and one output y. The inputs and outputs x_1, y_1, x_2, y_2 are all fuzzy variables, whose membership functions are given by

$$\mu_{x_1}(x) = \begin{cases} x - 1, & 1 \leq x < 1.6 \text{ or } 1.6 < x \leq 2 \\ 0.6, & x = 3 \\ 0, & \text{otherwise,} \end{cases}$$

$$\mu_{y_1}(x) = \begin{cases} x - 5, & 5 \leq x < 5.6 \text{ or } 5.6 < x \leq 6 \\ 0.6, & x = 3 \\ 0, & \text{otherwise,} \end{cases}$$

$$\mu_{x_2}(x) = \begin{cases} x - 3, & 3 \leq x < 3.6 \text{ or } 3.6 < x \leq 4 \\ 0.6, & x = 2 \\ 0, & \text{otherwise,} \end{cases}$$

$$\mu_{y_2}(x) = \begin{cases} x - 3, & 3 \leq x < 3.6 \text{ or } 3.6 < x \leq 4 \\ 0.6, & x = 5 \\ 0, & \text{otherwise.} \end{cases}$$

4.2 Fuzzy DEA Models

Table 4.1 The results of using credibility measure

DMU_i	Total slacks	Evaluation result
DMU_1	0	Efficient
DMU_2	3.6	Inefficient

Table 4.2 The results of using possibility measure

DMU_i	Total slacks	Evaluation result
DMU_1	3	Inefficient
DMU_2	4.8	Inefficient

As an example, we set the confidence level $\alpha = 0.6$. Table 4.1 shows the results when using model (4.2) to evaluate the two DMUs. It coincides with the fact that DMU_1 is efficient and DMU_2 is inefficient. When the credibility measure in model (4.2) is replaced by possibility measure, we can get the following possibility model:

$$\begin{cases} \max \quad \sum_{i=1}^{p} s_i^- + \sum_{j=1}^{q} s_j^+ \\ \text{subject to:} \\ \quad \text{Pos}\left\{\sum_{k=1}^{n} \tilde{x}_{ki}\lambda_k \leq \tilde{x}_{0i} - s_i^-\right\} \geq \alpha, \quad i = 1, 2, \cdots, p \\ \quad \text{Pos}\left\{\sum_{k=1}^{n} \tilde{y}_{kj}\lambda_k \geq \tilde{y}_{0j} + s_j^+\right\} \geq \alpha, \quad j = 1, 2 \cdots, q \\ \quad \sum_{k=1}^{n} \lambda_k = 1 \\ \quad \lambda_k \geq 0, \quad k = 1, 2, \cdots, n \\ \quad s_i^- \geq 0, \quad i = 1, 2 \cdots, p \\ \quad s_j^+ \geq 0, \quad j = 1, 2, \cdots, q. \end{cases} \quad (4.5)$$

Definition 4.3 (Possibility Efficiency). DMU_0 is possibility efficient if s_i^{-*} and s_j^{+*} are zero for $i = 1, 2, \cdots, p$ and $j = 1, 2 \cdots, q$, where s_i^{-*} and s_j^{+*} are optimal solutions of (4.5).

After evaluating by model (4.5), we can get the results in Table 4.2. It does not coincide with the fact that both DMU_1 and DMU_2 are inefficient. The reason comes from that the possibility measure has no self-duality property. Thus, this chapter uses credibility measure to model the fuzzy DEA.

Example 4.2. This numerical example is presented to illustrate the fuzzy DEA model (4.2). Table 4.3 provides the information of all the DMUs. There are two fuzzy inputs and two fuzzy outputs, which are all trapezoidal fuzzy variables.

Table 4.4 shows the results of evaluating DMUs with confidence level $\alpha = 0.7$. The results can be interpreted in the following way: DMU_1 and DMU_3 are inefficient, whereas DMU_2, DMU_4, and DMU_5 are efficient. Moreover, DMU_3 is more efficient than DMU_1 from the total slacks $\sum_{i=1}^{p} s_i^{-*} + \sum_{j=1}^{q} s_j^{+*}$, since they are both inefficient.

Table 4.3 DMUs with two fuzzy inputs and two fuzzy outputs

DMU_i	Input 1	Input 2	Output 1	Output 2
1	(44,47,50,53)	(24,26,28,30)	(25,26,27,28)	(36,38,41,44)
2	(23,25,27,29)	(17,19,21,23)	(24,25,26,27)	(41,45,49,51)
3	(45,48,51,54)	(29,32,35,38)	(27,32,37,42)	(35,40,45,50)
4	(35,39,43,47)	(19,21,23,25)	(38,40,42,44)	(36,38,40,42)
5	(61,67,73,75)	(31,36,41,46)	(42,47,52,58)	(71,74,77,80)

Table 4.4 Results of evaluating the DMUs with $\alpha = 0.7$

DMUs	$(\lambda_1^*, \lambda_2^*, \lambda_3^*, \lambda_4^*, \lambda_5^*)$	$\sum_{i=1}^{p} s_i^{-*} + \sum_{j=1}^{q} s_j^{+*}$	The result of evaluating
DMU_1	(0.0, 0.8, 0.0, 0.2, 0.0)	17.8	Inefficiency
DMU_2	(0.0, 1.0, 0.0, 0.0, 0.0)	0	Efficiency
DMU_3	(0.0, 0.0, 0.0, 1.0, 0.0)	9.6	Inefficiency
DMU_4	(0.0, 0.0, 0.0, 1.0, 0.0)	0	Efficiency
DMU_5	(0.0, 0.0, 0.0, 0.0, 1.0)	0	Efficiency

Table 4.5 Results of evaluating the DMUs with different confidence level α

α	DMU_1	DMU_2	DMU_3	DMU_4	DMU_5
0.5	Inefficiency	Efficiency	Inefficiency	Efficiency	Efficiency
0.6	Inefficiency	Efficiency	Inefficiency	Efficiency	Efficiency
0.7	Inefficiency	Efficiency	Inefficiency	Efficiency	Efficiency
0.8	Inefficiency	Efficiency	Efficiency	Efficiency	Efficiency
0.9	Inefficiency	Efficiency	Efficiency	Efficiency	Efficiency

Fuzzy efficiencies obtained from the model (4.2) for different confidence levels α are shown in Table 4.5. DMU_1 is inefficient at all confidence levels, whereas DMU_2, DMU_4, and DMU_5 are always efficient at all levels. The result that DMU_3 is efficient at higher levels and inefficient at lower levels coincides with Theorem 4.2. It can be seen that the number of the efficient DMUs is affected by the confidence level α. The higher the confidence level α is, the more the number of efficient DMUs is. These phenomena indicate that fuzzy DEA is more complex than the traditional DEA because of the inherent fuzziness contained in inputs and outputs.

4.3 Sensitivity and Stability

During the recent years, the issue of sensitivity and stability of DEA has been extensively studied. The first DEA sensitivity analysis paper by Charnes et al. [5] examined change in a single output. This is followed by a series of sensitivity analysis articles. Charnes and Neralic [3] gave some sufficient conditions in preserving efficiency. Charnes et al. [6,7] developed a sensitivity analysis technique on super-efficiency DEA model for the situation where simultaneous proportional

4.3 Sensitivity and Stability

change is assumed in all inputs and outputs for a specific DMU under consideration. This data variation condition is relaxed in Zhu [47] and Seiford and Zhu [40] to a situation, where inputs or outputs can be changed individually and the largest stability region is obtained. The DEA sensitivity analysis methods we have just reviewed are all developed for the situation, where data variations are only applied to the efficient DMU under evaluation and the data for the remaining DMUs are assumed to be fixed. Obviously, this assumption may not be realistic, since possible data errors may occur in each DMU. Seiford and Zhu [41] generalized the technique in Zhu [47] and Seiford [39] to the case, where the efficiency of the under evaluation efficient DMU is deteriorating while the efficiencies of the other DMUs are improving. This section will consider the sensitivity and stability analysis in fuzzy DEA. The focus is on the stability of classification of DMUs (decision-making units) into efficient and inefficient performers.

Theorem 4.3 (Wen et al. [45]). *If DMU_0 is α-inefficient, then the optimal solution satisfying $\lambda_0^*(\alpha) = 0$.*

Proof. For a fixed α, suppose the optimal solution is $\left(\lambda_j^*, \lambda_0^*, s_i^{-*}, s_j^{+*}\right)$ and the optimal objective value is $\sum_{i=1}^{p} s_i^{-*} + \sum_{j=1}^{q} s_j^{+*}$. If $\lambda_0^* = 0$, then the Theorem has been proved. Otherwise, let $\lambda_0 > 0$. Since DMU_0 is inefficient, there exists at least one $s_i^{-*} > 0$ or $s_j^{+*} > 0$, $i = 1, 2, \cdots, p$, $j = 1, 2, \cdots, q$. Without loss of generality, we assume $s_1^{-*} > 0$. If $\lambda_0^* = 1$, then $\text{Cr}\left\{\tilde{x}_{01} \leq \tilde{x}_{01} - s_1^{-*}\right\} = 0$. The contradiction implies that $\lambda_0^* \neq 1$.

$$\text{Cr}\left\{\sum_{k=1}^{n} \tilde{x}_{ki} \lambda_k^* \leq \tilde{x}_{0i} - s_i^{-*}\right\}$$

$$= \text{Cr}\left\{\sum_{k=1, k \neq 0}^{n} \tilde{x}_{ki} \lambda_k^* + \tilde{x}_{0i} \lambda_0^* \leq \tilde{x}_{0i} - s_i^{-*}\right\}$$

$$= \text{Cr}\left\{\frac{(1-\lambda_0^*)\left(\sum_{k=1, k \neq 0}^{n} \tilde{x}_{ki} \lambda_k^*\right)}{(1-\lambda_0^*)} \leq (1-\lambda_0^*) \tilde{x}_{0i} - s_i^{-*}\right\}$$

$$= \text{Cr}\left\{\frac{\sum_{k=1, k \neq 0}^{n} \tilde{x}_{ki} \lambda_k^*}{1-\lambda_0^*} \leq \tilde{x}_{0i} - \frac{s_i^{-*}}{(1-\lambda_0^*)}\right\}$$

$$\geq \alpha, \ i = 1, 2, \cdots, p.$$

Similarly we can get

$$\text{Cr}\left\{\sum_{k=1}^{n} \tilde{y}_{kj}\lambda_k^* \geq \tilde{y}_{0j} + s_j^{+*}\right\}$$

$$= \text{Cr}\left\{\frac{\sum_{k=1,k\neq 0}^{n} \tilde{y}_{kj}\lambda_k^*}{1-\lambda_0^*} \geq \tilde{y}_{0j} + \frac{s_j^{+*}}{(1-\lambda_0^*)}\right\}$$

$$\geq \alpha, \; j = 1, 2, \cdots, q.$$

It is easy to prove that $\dfrac{\sum_{k=1,k\neq 0}^{n} \lambda_k^*}{1-\lambda_0^*} = 1$. Thus, $\left(\dfrac{\lambda_1^*}{\sum_{k=1,k\neq 0}^{n} \lambda_k^*}, \cdots, \dfrac{\lambda_{0-1}^*}{\sum_{k=1,k\neq 0}^{n} \lambda_k^*},\right.$

$\left.0, \dfrac{\lambda_{0+1}^*}{\sum_{k=1,k\neq 0}^{n} \lambda_k^*}, \cdots, \dfrac{\lambda_n^*}{\sum_{k=1,k\neq 0}^{n} \lambda_k^*}\right)$ is a feasible solution and the objective value is

$\dfrac{1}{1-\lambda_0^*}\left(\sum_{i=1}^{p} s_i^{-*} + \sum_{j=1}^{q} s_j^{+*}\right) > \sum_{i=1}^{p} s_i^{-*} + \sum_{j=1}^{q} s_j^{+*}$, since $0 < \lambda_0^* < 1$, which leads to a contradiction with the assumption. Thus, $\lambda_0^* = 0$.

Theorem 4.4 (Wen et al. [44]). *An α-inefficient DMU_0 becomes α-efficient if $(\hat{x}_0, \hat{y}_0) = (\tilde{x}_0 - s^{-*}(\alpha), \tilde{y}_0 + s^{+*}(\alpha))$, in which $s^{-*}(\alpha)$ and $s^{+*}(\alpha)$ are optimal solution of (4.2).*

Proof. The efficiency of $(\hat{x}_0, \hat{y}_0)$ is evaluated by solving the problem below:

$$\begin{cases} \max \; \sum_{i=1}^{p} s_i^- + \sum_{j=1}^{q} s_j^+ \\ \text{subject to:} \\ \quad \text{Cr}\left\{\sum_{k=1}^{n} \tilde{x}_{ki}\lambda_k \leq \hat{x}_{0i} - s_i^-\right\} \geq \alpha, \; i = 1, 2\cdots, p \\ \quad \text{Cr}\left\{\sum_{k=1}^{n} \tilde{y}_{kj}\lambda_k \geq \hat{y}_{0j} + s_j^+\right\} \geq \alpha, \; j = 1, 2, \cdots, q \\ \quad \sum_{k=1}^{n} \lambda_k = 1 \\ \quad \lambda_k \geq 0, \; k = 1, 2, \cdots, n \\ \quad s_i^- \geq 0, \; i = 1, 2, \cdots, p \\ \quad s_j^+ \geq 0, \; j = 1, 2\cdots, q. \end{cases} \quad (4.6)$$

4.3 Sensitivity and Stability

Let an optimal solution be $(\hat{\lambda}, \hat{s}^+, \hat{s}^-)$. By inserting the formulae $(\hat{x}_0, \hat{y}_0) = (\tilde{x}_0 - s^{-*}, \tilde{y}_0 + s^{+*})$ into constraints, we have

$$\text{Cr}\left\{\sum_{k=1}^{n} \tilde{x}_{ki} \hat{\lambda}_k \leq \tilde{x}_{0i} - \hat{s}_i^- - s_i^{-*}\right\} \geq \alpha, \quad i = 1, 2 \cdots, p,$$

$$\text{Cr}\left\{\sum_{k=1}^{n} \tilde{y}_{kj} \hat{\lambda}_k \geq \tilde{y}_{0j} + \hat{s}_j^+ + s_j^{+*}\right\} \geq \alpha, \quad j = 1, 2, \cdots, q.$$

Now we can also write the constraints as

$$\text{Cr}\left\{\sum_{k=1}^{n} \tilde{x}_{ki} \hat{\lambda}_k \leq \tilde{x}_{0i} - \tilde{s}_i^-\right\} \geq \alpha, \quad i = 1, 2 \cdots, p,$$

$$\text{Cr}\left\{\sum_{k=1}^{n} \tilde{y}_{kj} \hat{\lambda}_k \geq \tilde{y}_{0j} + \tilde{s}_j^+\right\} \geq \alpha, \quad j = 1, 2, \cdots, q$$

where $\tilde{s}^+ = \hat{s}^+ + s^{+*} \geq 0$ and $\tilde{s}^- = \hat{s}^- + s^{-*} \geq 0$. Furthermore, we have

$$\sum_{i=1}^{p} \tilde{s}_i^- + \sum_{j=1}^{q} \tilde{s}_j^+ = \left(\sum_{i=1}^{p} \hat{s}_i^- + s_i^{-*}\right) + \left(\sum_{j=1}^{q} \hat{s}_j^+ + s_j^{+*}\right) \leq \sum_{i=1}^{p} s_i^{-*} \sum_{j=1}^{q} s_j^{+*}$$

since $\sum_{i=1}^{p} s_i^{-*} + \sum_{j=1}^{q} s_j^{+*}$ is maximal. It follows that we have $\sum_{i=1}^{p} \hat{s}_i^- + \sum_{j=1}^{q} \hat{s}_j^+ = 0$ which implies that all components $\hat{s}_i^-$ and $\hat{s}_j^+$ are zero. Hence, fuzzy efficiency is achieved as claimed.

Theorem 4.4 has given the stable region when the DMU is inefficient. But we also want to know the efficient radius of efficient DMUs. For this purpose, the following model is proposed:

$$\begin{cases} \min \sum_{i=1}^{p} t_i^+ + \sum_{j=1}^{q} t_j^- \\ \text{subject to:} \\ \quad \text{Cr}\left\{\sum_{k=1, k \neq 0}^{n} \tilde{x}_{ki} \lambda_k \leq \tilde{x}_{0i} + t_i^+\right\} \geq \alpha, \quad i = 1, 2, \cdots, p \\ \quad \text{Cr}\left\{\sum_{k=1, k \neq 0}^{n} \tilde{y}_{kj} \lambda_k \geq \tilde{y}_{0j} - t_j^-\right\} \geq \alpha, \quad j = 1, 2, \cdots, q \\ \quad \sum_{k=1}^{n} \lambda_k = 1 \\ \quad \lambda_k \geq 0, \quad k = 1, 2, \cdots, n \\ \quad t_i^+ \geq 0, \quad i = 1, 2, \cdots, p \\ \quad t_j^- \geq 0, \quad j = 1, 2, \cdots, q. \end{cases} \quad (4.7)$$

Theorem 4.5 (Wen et al. [45]). *The α-efficient DMU_0 stays α-efficient if $(\hat{x}_0, \hat{y}_0) = (\tilde{x}_0 + t^{+*}(\alpha), \tilde{y}_0 - t^{-*}(\alpha))$, where $t^{+*}(\alpha)$ and $t^{-*}(\alpha)$ are optimal solution of (5.10).*

Proof. Consider the following DEA model for evaluating the relative efficiency of the adjusted DMU_0:

$$\begin{cases} \max \quad \sum_{i=1}^{p} t_i^+ + \sum_{j=1}^{q} t_j^- \\ \text{subject to:} \\ \operatorname{Cr}\left\{ \sum_{k=1, k \neq 0}^{n} \tilde{x}_{ki} \lambda_k + (\tilde{x}_{0i} + t_i^{+*}) \lambda_0 \leq (\tilde{x}_{0i} + t_i^{+*}) - s_i^- \right\} \geq \alpha, \\ \qquad\qquad\qquad\qquad\qquad\qquad\qquad\qquad\qquad i = 1, 2, \cdots, p \\ \operatorname{Cr}\left\{ \sum_{k=1, k \neq 0}^{n} \widetilde{yj}_{kj} \lambda_k + (\tilde{y}_0 - t_j^{-*}) \lambda_0 \geq (\widetilde{yj}_0 - t_j^{-*}) + s_j^+ \right\} \geq \alpha, \\ \qquad\qquad\qquad\qquad\qquad\qquad\qquad\qquad\qquad j = 1, 2, \cdots, q \\ \sum_{k=1}^{n} \lambda_k = 1 \\ \lambda_k \geq 0, \quad k = 1, 2, \cdots, n \\ s_i^- \geq 0, \quad i = 1, 2, \cdots, p \\ s_j^+ \geq 0, \quad j = 1, 2, \cdots, q. \end{cases} \quad (4.8)$$

For a fixed α, let the optimal solution to be $(\lambda_j^*, \lambda_0^*, s^{-*}, s^{+*})$ and assume that the DMU_0 is inefficient. From Theorem 4.3, we get $\lambda_0^* = 0$. Thus, this optimal solution is a feasible solution for (5.10). Hence, $t_i^{+*} - s_i^{-*} \geq t_i^{+*}$ and $t_j^{-*} - s_j^{+*} \geq t_j^{-*}$, which means $s_i^{-*} = 0$ and s_j^{+*}, $i = 1, 2, \cdots, p$, $j = 1, 2, \cdots, q$. This leads to a contradiction with the assumption.

From the above analysis, the ranges of inputs and outputs and radius of stability of DMU_0 are identified as follows:

1. If DMU_0 is α-inefficient by solving model (4.2), then DMU_0 stays α-inefficient if $(\hat{x}_0, \hat{y}_0) = (\tilde{x}_0 - s^-, \tilde{y}_0 + s^+)$, in which $s^- = \left\{ (s_1^-, \cdots, s_p^-) \mid 0 \leq s_i^- < s_i^{-*}(\alpha), \ i = 1, 2, \cdots, p \right\}$, $s^+ = \left\{ (s_1^+, \cdots, s_q^+) \mid 0 \leq s_j^+ < s_j^{+*}(\alpha), \ j = 1, 2, \cdots, q \right\}$, where $s_i^{-*}(\alpha)$ and $s_j^{+*}(\alpha)$ are optimal solutions of (4.2).
2. If DMU_0 is α-efficient by solving model (4.2), then we use model (4.7) to account the efficient radius. DMU_0 stays α-efficient if $(\hat{x}_0, \hat{y}_0) = (\tilde{x}_0 + t^+, \tilde{y}_0 - t^-)$, in which $t^+ = \left\{ (t_1, \cdots, t_p) \mid 0 \leq t_i \leq t_i^{+*}(\alpha), \ i = 1, 2, \cdots, p \right\}$ and $t^- = \left\{ (t_1, \cdots, t_q) \mid 0 \leq t_j \leq t_j^{-*}(\alpha), \ j = 1, 2, \cdots, q \right\}$, where $t_i^{+*}(\alpha)$ and $t_j^{-*}(\alpha)$ are optimal solutions of (4.7).

4.3 Sensitivity and Stability

From the properties of the α-optimistic and α-pessimistic values, we can rewrite the fuzzy DEA model (4.7) as follows:

$$\begin{cases} \min \quad \sum_{i=1}^{p} t_i^+ + \sum_{j=1}^{q} t_j^- \\ \text{subject to:} \\ \quad \sum_{k=1,k\neq 0}^{n} \lambda_k (\tilde{x}_{ki})_{\inf}^{\alpha} \leq (\tilde{x}_{0i})_{\sup}^{\alpha} + t_i^+, \quad i = 1,2,\cdots,p \\ \quad \sum_{k=1,k\neq 0}^{n} \lambda_k (\tilde{y}_{kj})_{\sup}^{\alpha} \geq (\tilde{y}_{0j})_{\inf}^{\alpha} - t_j^-, \quad j = 1,2,\cdots,q \\ \quad \sum_{k=1}^{n} \lambda_k = 1 \\ \quad \lambda_k \geq 0, \quad k = 1,2,\cdots,n \\ \quad t_i^+ \geq 0, \quad i = 1,2,\cdots,p \\ \quad t_j^- \geq 0, \quad j = 1,2,\cdots,q \end{cases} \quad (4.9)$$

which is a linear programming model. Thus, they can be easily solved by many traditional methods.

When the inputs and outputs are trapezoidal fuzzy variables, following Liu's book [37], the fuzzy model (4.9) becomes the following linear programming:

$$\begin{cases} \min \quad \sum_{i=1}^{p} t_i^+ + \sum_{j=1}^{q} t_j^- \\ \text{subject to:} \\ \quad \sum_{k=1,k\neq 0}^{n} \lambda_k \left[2(1-\alpha)x_{ki}^c + (2\alpha-1)x_{ki}^d \right] \leq 2(1-\alpha)x_{0i}^b + (2\alpha-1)x_{0i}^a + t_i^-, \\ \qquad\qquad\qquad\qquad\qquad\qquad\qquad\qquad i = 1,2\cdots,p \\ \quad \sum_{k=1,k\neq 0}^{n} \lambda_k \left[2(1-\alpha)y_{kj}^b + (2\alpha-1)y_{kj}^a \right] \geq 2(1-\alpha)y_{0j}^c + (2\alpha-1)y_{0j}^d - t_j^+, \\ \qquad\qquad\qquad\qquad\qquad\qquad\qquad\qquad j = 1,2,\cdots,q \\ \quad \sum_{j=1}^{n} \lambda_k = 1 \\ \quad \lambda_k \geq 0, \quad k = 1,2,\cdots,n \\ \quad t_i^- \geq 0, \quad i = 1,2,\cdots,p \\ \quad t_j^+ \geq 0, \quad j = 1,2\cdots,q \end{cases}$$

(4.10)

in which the trapezoidal fuzzy variable $\left(x_{ki}^a, x_{ki}^b, x_{ki}^c, x_{ki}^d\right)$ represents the ith input of DMU_k and $\left(y_{kj}^a, y_{kj}^b, y_{kj}^c, y_{kj}^d\right)$ represents the ith output of DMU_k, respectively.

Example 4.3. This numerical example is presented to illustrate the sensitivity and stability of the DMUs. In this example, we also use the data in Table 4.3. From Table 4.4, we get that DMU_1 and DMU_3 are inefficient, whereas DMU_2, DMU_4, and DMU_5 are efficient, when the confidence level $\alpha = 0.7$.

Table 4.6 Sensitivity analysis results of inefficient DMUs by model (4.2)

DMUs	s_1^{-*}	s_2^{-*}	s_1^{+*}	s_2^{+*}
DMU$_1$	14.78	3.01	0.00	0.01
DMU$_3$	2.20	7.00	0.20	0.20

Table 4.7 Sensitivity analysis results of efficient DMUs by model (4.7)

DMUs	s_1^{-*}	s_2^{-*}	s_1^{+*}	s_2^{+*}
DMU$_2$	9.00	0.00	0.00	20.58
DMU$_4$	0.00	1.60	18.20	0.00
DMU$_5$	0.00	0.00	22.77	32.12

Table 4.6 reports the sensitivity analysis results for inefficient DMUs by model (4.2). In Table 4.6, the columns 2 and 3 report lower bounds of variation ranges of inputs, and the columns 4 and 5 are upper bounds of variation ranges of outputs. For instance, let us consider DMU$_1$ in Table 4.3. It stays inefficient when $(\hat{x}_{11}, \hat{x}_{12}, \hat{y}_{11}, \hat{y}_{12}) = (\tilde{x}_{11} - r_{x1}, \tilde{x}_{12} - r_{x2}, \tilde{y}_{11}, \tilde{y}_{12} + r_{y2})$, in which $0 \leq r_{x1} < 14.78$, $0 \leq r_{x2} < 3.01$ and $0 \leq r_{y2} < 0.01$.

Table 4.7 reports the sensitivity analysis results for efficient DMUs by model (4.7). In Table 4.7, the columns 2 and 3 report upper bounds of variation ranges of inputs, and the columns 4 and 5 are lower bounds of variation ranges of outputs. For instance, let us consider DMU$_4$ in Table 4.7. It stays efficient when $(\hat{x}_{41}, \hat{x}_{42}, \hat{y}_{41}, \hat{y}_{42}) = (\tilde{x}_{41}, \tilde{x}_{42} + r_{x2}, \tilde{y}_{41} - r_{y1}, \tilde{y}_{42})$, in which $0 \leq r_{x2} \leq 1.60$ and $0 \leq r_{y1} \leq 18.20$.

The similar interpretation can be stated for other rows in Tables 4.6 and 4.7.

4.4 Fuzzy DEA Ranking Criteria

This section will introduce some fully ranking methods in fuzzy DEA. Four types of fuzzy DEA fully ranking criteria are to be investigated.

4.4.1 Expected Ranking Criterion

The essential idea of the fuzzy expected DEA model is to optimize the expected value of $\dfrac{v^T \tilde{y}_0}{u^T \tilde{x}_0}$ subject to some chance constraints, then we have the first type of the fuzzy DEA model:

$$\begin{cases} \theta = \max_{u,v} E\left[\dfrac{v^T \tilde{y}_0}{u^T \tilde{x}_0}\right] \\ \text{subject to:} \\ \quad \text{Cr}\left\{v^T \tilde{y}_k \leq u^T \tilde{x}_k\right\} \geq \alpha, \ k = 1, 2, \cdots, n \\ \quad u \geq 0 \\ \quad v \geq 0 \end{cases} \quad (4.11)$$

in which $\alpha \in (0.5, 1]$.

4.4 Fuzzy DEA Ranking Criteria

Definition 4.4. A vector $(u, v) \geq 0$ is called a feasible solution to the fuzzy programming model (4.11) if

$$\mathrm{Cr}\left\{v^T \tilde{y}_k \leq u^T \tilde{x}_k\right\} \geq \alpha \tag{4.12}$$

for $k = 1, 2, \cdots, n$.

Definition 4.5. A feasible solution (u^*, v^*) is called an expected optimal solution to the fuzzy programming model (4.11) if

$$E\left[\frac{v^{*T} \tilde{y}_0}{u^{*T} \tilde{x}_0}\right] \geq E\left[\frac{v^T \tilde{y}_0}{u^T \tilde{x}_0}\right] \tag{4.13}$$

for any feasible solution (u, v).

Expected Ranking Criterion: The greater the optimal objective value is, the more efficient DMU$_0$ is ranked.

4.4.2 Optimistic Ranking Criterion

Chance-constrained programming (CCP), which was initialized by Charnes and Cooper [2], offers a powerful means for modeling stochastic decision systems. The essential idea of chance-constrained programming is to optimize some critical value with a given confidence level subject to some chance constraints. Inspired by this idea, Wen et al. [43] extended it to fuzzy DEA models. Assuming that the decision makers want to maximize the optimistic value of the uncertain objective at a given confidence level, we have the second type of fuzzy DEA model:

$$\begin{cases} \max \overline{f} \\ \text{subject to:} \\ \quad \mathrm{Cr}\left\{\dfrac{v^T \tilde{y}_0}{u^T \tilde{x}_0} \leq \overline{f}\right\} \leq \alpha \\ \quad \mathrm{Cr}\left\{v^T \tilde{y}_j \leq u^T \tilde{x}_j\right\} \geq \alpha, \; j = 1, 2, \cdots, n \\ \quad u \geq 0 \\ \quad v \geq 0 \end{cases} \tag{4.14}$$

in which $\alpha \in [0.5, 1]$.

Definition 4.6 (Wen et al. [43]). Let ξ and η be fuzzy variables. We say $\xi > \eta$ if and only if, for some predetermined confidence level $\alpha \in (0, 1]$, we have

$$\sup\left\{r \mid \mathrm{Cr}\{\xi \leq r\} \leq \alpha\right\} > \sup\left\{r \mid \mathrm{Cr}\{\eta \leq r\} \leq \alpha\right\}. \tag{4.15}$$

Ranking Criterion: The greater the optimal objective is, the more efficient DMU$_0$ is ranked.

We have given the fuzzy DEA model, but the model is too complex to solve by traditional methods. This section aims to make the model easy to compute when the membership function of the fuzzy inputs and outputs have some particular characters.

Definition 4.7 (Wen et al. [43]). A real-valued function f defined on a convex set $X \in \mathcal{R}^n$ is said to be quasiconcave if

$$f(\lambda x + (1-\lambda)y) \geq \min\{f(x), f(y)\}$$

for any $x, y \in X$ and $0 < \lambda < 1$.

Theorem 4.6. *Suppose that ξ_1, ξ_2 are fuzzy variables defined on credibility space $(\Theta, \mathcal{P}(\Theta), \text{Cr})$. If $\text{Cr}\{\xi_1 = x\}$ and $\text{Cr}\{\xi_2 = x\}$ are quasiconcave, then for any given $0.5 \leq \alpha \leq 1$:*

$\text{Cr}\{\xi_1 + \xi_2 \leq b\} \geq \alpha$ *if and only if* $(\xi_1)_{1-\alpha}^U + (\xi_2)_{1-\alpha}^U \leq b$.
$\text{Cr}\{\xi_1 + \xi_2 \leq b\} \leq \alpha$ *if and only if* $(\xi_1)_{1-\alpha}^U + (\xi_2)_{1-\alpha}^U \geq b$.

Proof. If $\text{Cr}\{\xi_1 + \xi_2 \leq b\} \geq \alpha$, $0.5 \leq \alpha \leq 1$, then we have

$$\text{Cr}\{\xi_1 + \xi_2 \leq b\}$$
$$= 1 - \sup_{x_1+x_2>b} \{\text{Cr}\{\xi_1 = x_1\} \wedge \text{Cr}\{\xi_2 = x_2\}\}$$
$$\geq \alpha.$$

Hence,

$$\sup_{x_1+x_2>b} \{\text{Cr}\{\xi_1 = x_1\} \wedge \text{Cr}\{\xi_2 = x_2\}\} \leq 1 - \alpha.$$

Suppose that

$$\left(x_1^*, x_2^*\right) = \arg \sup_{x_1, x_2 \in R} \{\text{Cr}\{\xi_1 = x_1\} \wedge \text{Cr}\{\xi_2 = x_2\} | x_1 + x_2 > b\} \leq 1 - \alpha\}. \tag{4.16}$$

It follows that $\text{Cr}\{\xi_1 = x_1^*\} \wedge \text{Cr}\{\xi_2 = x_2^*\} \leq 1 - \alpha$ and $x_1^* + x_2^* > b$.
Since $\text{Cr}\{\xi_1 = x_1^*\} \wedge \text{Cr}\{\xi_2 = x_2^*\} \leq 1 - \alpha$ implies that $\text{Cr}\{\xi_1 = x_1^*\} \leq 1 - \alpha$, $\text{Cr}\{\xi_2 = x_2^*\} \leq 1 - \alpha$. From that the functions $\text{Cr}\{\xi_1 = x_1\}$ and $\text{Cr}\{\xi_2 = x_2\}$ are quasiconcave, we have

$$x_1^* \geq (\xi_1)_{1-\alpha}^U, \quad x_2^* \geq (\xi_2)_{1-\alpha}^U.$$

4.4 Fuzzy DEA Ranking Criteria

Then we get $(\xi_1)_{1-\alpha}^U + (\xi_2)_{1-\alpha}^U \leq b$. Otherwise,

$$(\xi_1)_{1-\alpha}^U + (\xi_2)_{1-\alpha}^U > b,$$
$$\mathrm{Cr}\{\xi_1 = (\xi_1)_{1-\alpha}^U\} \geq \mathrm{Cr}\{\xi_1 = x_1^*\},$$
$$\mathrm{Cr}\{\xi_2 = (\xi_2)_{1-\alpha}^U\} \geq \mathrm{Cr}\{\xi_2 = x_2^*\}$$

which contradicts with (4.16).

Conversely, if $(\xi_1)_{1-\alpha}^U + (\xi_2)_{1-\alpha}^U \leq b$, we get

$$\mathrm{Cr}\{\xi_1 = a_1\} \leq 1 - \alpha, \; \mathrm{Cr}\{\xi_2 = a_1\} \leq 1 - \alpha$$

since $a_1 > (\xi_1)_{1-\alpha}^U$ and $a_2 > (\xi_2)_{1-\alpha}^U$.

Consequently, $\sup\limits_{x_1, x_2 \in R} \{\mathrm{Cr}\{\xi_1 = x_1\} \wedge \mathrm{Cr}\{\xi_2 = x_2\}\} | x_1 + x_2 > b\} \leq 1 - \alpha.$

Then

$$\mathrm{Cr}\{\xi_1 + \xi_2 \leq b\}$$
$$= 1 - \sup_{x_1 + x_2 > b} \{\mathrm{Cr}\{\xi_1 = x_1\} \wedge \mathrm{Cr}\{\xi_2 = x_2\}\}$$
$$\geq \alpha.$$

Theorem 4.7. *Let $(\xi)_\alpha^L$ and $(\xi)_\alpha^U$ be the lower and upper bounds of α-level set of ξ, respectively. Then we have:*

(a) *If $c \geq 0$, then $(c\xi)_\alpha^U = c(\xi)_\alpha^U$ and $(c\xi)_\alpha^L = c(\xi)_\alpha^L$.*
(b) *If $c \leq 0$, then $(c\xi)_\alpha^U = c(\xi)_\alpha^L$ and $(c\xi)_\alpha^L = c(\xi)_\alpha^U$.*

Proof. (a) If $c \geq 0$, then the part (a) is obviously valid. When $c > 0$, we have

$$(c\xi)_\alpha^U = \sup\{x | \mu_{c\xi}(x) \geq \alpha\}$$
$$= c \sup\left\{\frac{x}{c} | \mu_\xi(\frac{x}{c}) \geq \alpha\right\}$$
$$= c(\xi)_\alpha^U$$

A similar way may prove that $(c\xi)_\alpha^L = c(\xi)_\alpha^L$.

In order to prove the part (b), it suffices to verify that $(-\xi)_\alpha^U = -(\xi)_\alpha^L$ and $(-\xi)_\alpha^L = -(\xi)_\alpha^U$. In fact, for any $\alpha \in (0, 1]$, we have

$$(-\xi)_\alpha^U = \sup\{x | \mu_{-\xi}(x) \geq \alpha\}$$
$$= -\inf\{-x | \mu_\xi(-x) \geq \alpha\}$$
$$= -(\xi)_\alpha^L$$

Similarly, we may prove that $(-\xi)_\alpha^L = -(\xi)_\alpha^U$.

Following Theorems 4.6 and 4.7, the fuzzy DEA model (4.14) becomes

$$\begin{cases} \max \overline{f} \\ \text{subject to:} \\ \quad \dfrac{v^T(\tilde{y}_0)^U_{2(1-\alpha)}}{u^T(\tilde{x}_0)^L_{2(1-\alpha)}} \geq \overline{f} \\ \quad v^T(\tilde{y}_j)^U_{2(1-\alpha)} - u^T(\tilde{x}_j)^L_{2(1-\alpha)} \leq 0, \ j = 1, 2, \cdots, n \\ \quad u \geq 0 \\ \quad v \geq 0 \end{cases} \quad (4.17)$$

in which $\alpha \in [0.5, 1]$. The model is equivalent to the fractional model as follows:

$$\begin{cases} \max \dfrac{v^T(\tilde{y}_0)^U_{2(1-\alpha)}}{u^T(\tilde{x}_0)^L_{2(1-\alpha)}} \\ \text{subject to:} \\ \quad v^T(\tilde{y}_j)^U_{2(1-\alpha)} - u^T(\tilde{x}_j)^L_{2(1-\alpha)} \leq 0, \ j = 1, 2, \cdots, n \\ \quad u \geq 0 \\ \quad v \geq 0 \end{cases} \quad (4.18)$$

which is equivalent to the linear programming model:

$$\begin{cases} \max \ v^T(\tilde{y}_0)^U_{2(1-\alpha)} \\ \text{subject to:} \\ \quad u^T(\tilde{x}_0)^L_{2(1-\alpha)} = 1 \\ \quad v^T(\tilde{y}_j)^U_{2(1-\alpha)} - u^T(\tilde{x}_j)^L_{2(1-\alpha)} \leq 0, \ j = 1, 2, \cdots, n \\ \quad u \geq 0 \\ \quad v \geq 0. \end{cases} \quad (4.19)$$

The model (4.19) can be easily solved by many methods, i.e., a standard LP solver [18], simplex algorithm [8], and so on.

When all the inputs and outputs are trapezoidal fuzzy variables, denoted by $x_{ik} = \left(x_{ik}^{r_1}, x_{ik}^{r_2}, x_{ik}^{r_3}, x_{ik}^{r_4}\right)$ and $y_{il} = \left(x_{il}^{r_1}, x_{il}^{r_2}, x_{il}^{r_3}, x_{il}^{r_4}\right)$, $i = 1, 2, \cdots, n$, $k = 1, 2, \cdots, p$, $l = 1, 2, \cdots, q$. Following Liu's book [37], the model (4.14) becomes the following linear programming:

$$\begin{cases} \max_{u,v} (2\alpha - 1)v^T y_0^{r_4} + 2(1-\alpha)v^T y_0^{r_3} \\ \text{subject to:} \\ \quad (2\alpha - 1)u^T x_0^{r_1} + 2(1-\alpha)u^T x_0^{r_2} = 1 \\ \quad (2\alpha - 1)\left(v^T y_j^{r_4} - u^T x_j^{r_1}\right) + 2(1-\alpha)\left(v^T y_j^{r_3} - u^T x_j^{r_2}\right) \leq 0, \\ \qquad\qquad\qquad\qquad\qquad\qquad\qquad\qquad j = 1, 2, \cdots, n \\ \quad u \geq 0 \\ \quad v \geq 0 \end{cases} \quad (4.20)$$

in which $\boldsymbol{x}_j^{rk} = \left(x_{j1}^{rk}, x_{j2}^{rk}, \cdots, x_{jp}^{rk}\right)$ and $\boldsymbol{y}_j^{rk} = \left(y_{j1}^{rk}, y_{j2}^{rk}, \cdots, y_{jp}^{rk}\right)$.

4.4.3 Maximal Chance Ranking Criterion

Sometimes the decision maker may want to maximize the chance of satisfying the event $\dfrac{\boldsymbol{v}^T \tilde{\boldsymbol{y}}_0}{\boldsymbol{u}^T \tilde{\boldsymbol{x}}_0} \geq 1$. In order to model this type of decision system, Liu [33, 34, 36] provided the dependent-chance programming (DCP). Wen and Li [42] applied the DCP into the DEA as follows:

$$\begin{cases} \max\limits_{u,v} \theta = \mathrm{Cr}\left\{\dfrac{\boldsymbol{v}^T \tilde{\boldsymbol{y}}_0}{\boldsymbol{u}^T \tilde{\boldsymbol{x}}_0} \geq 1\right\} \\ \text{subject to:} \\ \quad \mathrm{Cr}\left\{\boldsymbol{v}^T \tilde{\boldsymbol{y}}_k \leq \boldsymbol{u}^T \tilde{\boldsymbol{x}}_k\right\} \geq 1-\alpha,\ k = 1,2,\cdots,n \\ \quad \boldsymbol{u} \geq 0 \\ \quad \boldsymbol{v} \geq 0 \end{cases} \quad (4.21)$$

in which $\alpha \in (0, 0.5]$.

Definition 4.8 (Wen and Li [42]). A feasible solution $(\boldsymbol{u}^*, \boldsymbol{v}^*)$ is called an maximal chance optimal solution to the fuzzy programming model (4.21) if

$$\mathrm{Cr}\left\{\dfrac{\boldsymbol{v}^{*T} \tilde{\boldsymbol{y}}_0}{\boldsymbol{u}^{*T} \tilde{\boldsymbol{x}}_0} \geq 1\right\} \geq \mathrm{Cr}\left\{\dfrac{\boldsymbol{v}^T \tilde{\boldsymbol{y}}_0}{\boldsymbol{u}^T \tilde{\boldsymbol{x}}_0} \geq 1\right\} \quad (4.22)$$

for any feasible solution $(\boldsymbol{u}, \boldsymbol{v})$.

Ranking Method: The greater the optimal objective is, the more efficient DMU_0 is ranked.

When the inputs and outputs are trapezoidal fuzzy variables, we can easily get the next theorem by Liu's book [35]:

Theorem 4.8. Let ξ_i be independent trapezoidal fuzzy variables with $\xi_i = \left(r_1^i, r_2^i, r_3^i, r_4^i\right)$, $i = 1, 2, \cdots, n$; then

(a) $\mathrm{Cr}\left\{\sum\limits_{i=1}^{n} k_i \cdot \xi_i \leq b\right\} \geq \alpha$ if and only if $(2\alpha-1)\sum\limits_{i=1}^{n} k_i r_4^i + 2(1-\alpha)\sum\limits_{i=1}^{n} k_2 r_3^i \leq b$,

(b) $\mathrm{Cr}\left\{\sum\limits_{i=1}^{n} k_i \cdot \xi_i \geq b\right\} \geq \alpha$ if and only if $(2\alpha-1)\sum\limits_{i=1}^{n} k_i r_1^i + 2(1-\alpha)\sum\limits_{i=1}^{n} k_2 r_2^i \geq b$,

in which k_i are positive numbers and $0.5 \leq \alpha \leq 1$.

When all the inputs $\tilde{x}_i = (\tilde{x}_{i1}, \tilde{x}_{i2}, \cdots, \tilde{x}_{ip})$ and outputs $\tilde{y}_i = (\tilde{y}_{i1}, \tilde{y}_{i2}, \cdots, \tilde{y}_{iq})$ are trapezoidal fuzzy vectors, denoted by $\tilde{x}_{ik} = (\tilde{x}_{ik}^{r_1}, \tilde{x}_{ik}^{r_2}, \tilde{x}_{ik}^{r_3}, \tilde{x}_{ik}^{r_4})$ and $\tilde{y}_{il} = (\tilde{y}_{il}^{r_1}, \tilde{y}_{il}^{r_2}, \tilde{y}_{il}^{r_3}, \tilde{y}_{il}^{r_4})$, $i = 1, 2, \cdots, n$, $k = 1, 2, \cdots, p$, $l = 1, 2, \cdots, q$, the fuzzy DEA model (4.21) becomes the following fractional programming:

$$\begin{cases} \max_{u,v} \dfrac{v^T \tilde{y}_0^{r_4} - u^T \tilde{x}_0^{r_1}}{2\left(u^T \tilde{x}_0^{r_2} - v^T \tilde{y}_0^{r_3} - u^T \tilde{x}_0^{r_1} + v^T \tilde{y}_0^{r_4}\right)} \\ \text{subject to:} \\ (1 - 2\alpha)\left(v^T \tilde{y}_j^{r_4} - u^T \tilde{x}_j^{r_1}\right) + 2\alpha\left(v^T \tilde{y}_j^{r_3} - u^T \tilde{x}_j^{r_2}\right) \leq 0, \\ \qquad\qquad\qquad\qquad\qquad\qquad\qquad j = 1, 2, \cdots, n \\ u \geq 0 \\ v \geq 0 \end{cases} \quad (4.23)$$

which is equivalent to the following linear programming model:

$$\begin{cases} \max_{u,v} \quad v^T \tilde{y}_0^{r_4} - u^T \tilde{x}_0^{r_1} \\ \text{subject to:} \\ 2v^T\left(\tilde{y}_0^{r_4} - \tilde{y}_0^{r_3}\right) + 2u^T\left(\tilde{x}_0^{r_2} - \tilde{x}_0^{r_1}\right) = 1 \\ v^T\left[(1 - 2\alpha)\tilde{y}_j^{r_4} + 2\alpha \tilde{y}_j^{r_3}\right] - u^T\left[(1 - 2\alpha)\tilde{x}_j^{r_1} + 2\alpha \tilde{x}_j^{r_2}\right] \leq 0, \\ \qquad\qquad\qquad\qquad\qquad\qquad\qquad j = 1, 2, \cdots, n \\ u \geq 0 \\ v \geq 0, \end{cases} \quad (4.24)$$

in which $\tilde{x}_j^{r_k} = \left(x_{j1}^{r_k}, x_{j2}^{r_k}, \cdots, x_{jp}^{r_k}\right)$, $\tilde{y}_j^{r_k} = \left(y_{j1}^{r_k}, y_{j2}^{r_k}, \cdots, y_{jp}^{r_k}\right)$.

4.4.4 Hurwicz Ranking Criterion

Most of the DEA models evaluate the distance of DMU_0 to an efficient frontier, which can be considered an optimistic method in comparison. At the other hand, Jahanshahloo and Afzalinejad [28] proposed a ranking method which uses the distance to the inefficient frontier as its efficiency value. As a result, this section wants to rank all the DMUs by the Hurwicz criterion [26, 27], which attempts to find a middle ground between the optimistic and pessimistic criteria.

Similar to model (4.2), the pessimistic model, which considers the total distances to an inefficient frontier, can be given as

4.4 Fuzzy DEA Ranking Criteria

$$\begin{cases} \theta_2 = \max \sum_{i=1}^{p} s_i^- + \sum_{j=1}^{q} s_j^+ \\ \text{subject to:} \\ \quad \text{Cr}\left\{ \sum_{k=1}^{n} \tilde{x}_{ki} \lambda_k \geq \tilde{x}_{0i} + s_i^- \right\} \geq \alpha, \quad i = 1, 2, \cdots, p \\ \quad \text{Cr}\left\{ \sum_{k=1}^{n} \tilde{y}_{kj} \lambda_k \leq \tilde{y}_{0j} - s_j^+ \right\} \geq \alpha, \quad j = 1, 2 \cdots, q \\ \quad \sum_{j=1}^{n} \lambda_j = 1 \\ \quad \lambda_j \geq 0, \quad j = 1, 2, \cdots, n \\ \quad s_i^- \geq 0, \quad i = 1, 2, \cdots, p \\ \quad s_j^+ \geq 0, \quad j = 1, 2 \cdots, q. \end{cases} \quad (4.25)$$

Definition 4.9 (α-Inefficiency). DMU$_0$ is α-inefficient if s_i^{-*} and s_j^{+*} are zero for $i = 1, 2, \cdots, p$ and $j = 1, 2 \cdots, q$, where s_i^{-*} and s_j^{+*} are optimal solutions of (4.25).

Since $j = 0$ is one of the DMU$_j$, we can always get a solution with $\lambda_0 = 1, \lambda_j = 0$ ($j \neq 0$) and all slacks zero. Thus, fuzzy DEA models (4.2) and (4.25) have feasible solution and the optimal value $s_i^{-*} = s_j^{+*} = 0$ for all i, j.

Abovementioned two models are both extreme cases: One is too optimistic and the other is too pessimistic. Thus, we employ the Hurwicz criterion, suggested by Leonid Hurwicz [26,27] in 1951, which incorporates a measure of both by assigning a certain percentage weight λ to θ_1^* and $1 - \lambda$ to $-\theta_2^*$, $0 \leq \lambda \leq 1$:

$$\theta^* = \lambda \theta_1^* + (1 - \lambda)(-\theta_2^*) \quad (4.26)$$

which can be rewritten as

$$\theta^* = \lambda \theta_1^* - (1 - \lambda)\theta_2^*. \quad (4.27)$$

Ranking Criterion: The greater the value θ^* is, the less efficient DMU$_0$ is ranked.

In the Hurwicz criterion, the parameter $\lambda \in [0, 1]$, which reflects the degree of the decision maker's optimism, must be determined by the decision maker. Generally speaking, it is difficult to determine the appropriate λ for decision makers, since it varies from person to person. By varying the parameter λ, the Hurwicz criterion becomes various models, e.g., when $\lambda = 1$, the criterion is the traditional DEA model (4.2); when $\lambda = 0$, it degenerates to model (4.25). This fact means that the Hurwicz criterion is fairly flexible.

In some cases, $\theta_1^* = 0$ and $\theta_2^* = 0$, and then the ranking value $\theta^* = 0$. It says DMU_0 is both efficient and inefficient. This happens when DMU_0 is the best in some inputs and/or outputs while it is the worst in some other inputs and/or outputs. For example, if DMU_0 is the only DMU that has the largest value for input 1 and has the least amount of input 2, DMU_0 is both efficient and inefficient.

From the properties of the α-optimistic and α-pessimistic values, the fuzzy models (4.2) and (4.25) can be converted to the following linear programming models:

$$\begin{cases} \theta_1 = \max \sum_{i=1}^{p} s_i^- + \sum_{j=1}^{q} s_j^+ \\ \text{subject to:} \\ \sum_{k=1}^{n} \lambda_k (\tilde{x}_{ki})_{\inf}^\alpha + \lambda_0 \left[(\tilde{x}_{0i})_{\sup}^\alpha - (\tilde{x}_{0i})_{\inf}^\alpha \right] \leq (\tilde{x}_{0i})_{\sup}^\alpha - s_i^-, \quad i=1,2\cdots,p \\ \sum_{k=1}^{n} \lambda_k (\tilde{y}_{kj})_{\sup}^\alpha + \lambda_0 \left[(\tilde{y}_{0j})_{\inf}^\alpha - (\tilde{y}_{0j})_{\sup}^\alpha \right] \geq (\tilde{y}_{0j})_{\inf}^\alpha + s_j^+, \quad j=1,2,\cdots,q \\ \sum_{k=1}^{n} \lambda_k = 1 \\ \lambda_k \geq 0, \quad k=1,2,\cdots,n \\ s_i^- \geq 0, \quad i=1,2\cdots,p \\ s_j^+ \geq 0, \quad j=1,2,\cdots,q \end{cases}$$

(4.28)

and

$$\begin{cases} \theta_2 = \max \sum_{i=1}^{p} s_i^- + \sum_{j=1}^{q} s_j^+ \\ \text{subject to:} \\ \sum_{k=1}^{n} \lambda_k (\tilde{x}_{ki})_{\sup}^\alpha + \lambda_0 \left[(\tilde{x}_{0i})_{\inf}^\alpha - (\tilde{x}_{0i})_{\sup}^\alpha \right] \geq (\tilde{x}_{0i})_{\inf}^\alpha + s_i^-, \quad i=1,2\cdots,p \\ \sum_{k=1}^{n} \lambda_k (\tilde{y}_{kj})_{\inf}^\alpha + \lambda_0 \left[(\tilde{y}_{0j})_{\sup}^\alpha - (\tilde{y}_{0j})_{\inf}^\alpha \right] \leq (\tilde{y}_{0j})_{\sup}^\alpha - s_j^+, \quad j=1,2,\cdots,q \\ \sum_{k=1}^{n} \lambda_k = 1 \\ \lambda_k \geq 0, \quad k=1,2,\cdots,n \\ s_i^- \geq 0, \quad i=1,2\cdots,p \\ s_j^+ \geq 0, \quad j=1,2,\cdots,q. \end{cases}$$

(4.29)

When the inputs and outputs are trapezoidal fuzzy variables, following Liu's book [37], the fuzzy models become the following linear programming:

4.4 Fuzzy DEA Ranking Criteria

$$\begin{cases} \max \sum_{i=1}^{p} s_i^- + \sum_{j=1}^{q} s_j^+ \\ \text{subject to:} \\ \sum_{k=1}^{n} \lambda_k \left[2(1-\alpha)x_{ki}^c + (2\alpha-1)x_{ki}^d \right] \\ \quad + \lambda_0 \left[2(1-\alpha)\left(x_{0i}^b - x_{0i}^c\right) + (2\alpha-1)\left(x_{0i}^a - x_{0i}^d\right) \right] \\ \quad \leq 2(1-\alpha)x_{0i}^b + (2\alpha-1)x_{0i}^a - s_i^-, i = 1,2\cdots,p \\ \sum_{k=1}^{n} \lambda_k \left[2(1-\alpha)y_{kj}^b + (2\alpha-1)y_{kj}^a \right] \\ \quad + \lambda_0 \left[2(1-\alpha)\left(y_{0j}^c - y_{0j}^b\right) + (2\alpha-1)\left(y_{0j}^d - y_{0j}^a\right) \right] \\ \quad \geq 2(1-\alpha)y_{0j}^c + (2\alpha-1)y_{0j}^d + s_j^+, j = 1,2,\cdots,q \\ \sum_{j=1}^{n} \lambda_k = 1 \\ \lambda_k \geq 0, \quad k = 1,2,\cdots,n \\ s_i^- \geq 0, \quad i = 1,2,\cdots,p \\ s_j^+ \geq 0, \quad j = 1,2\cdots,q \end{cases} \quad (4.30)$$

and

$$\begin{cases} \max \sum_{i=1}^{p} s_i^- + \sum_{j=1}^{q} s_j^+ \\ \text{subject to:} \\ \sum_{k=1}^{n} \lambda_k \left[2(1-\alpha)x_{kj}^b + (2\alpha-1)x_{kj}^a \right] \\ \quad + \lambda_0 \left[2(1-\alpha)\left(x_{0i}^c - x_{0j}^b\right) + (2\alpha-1)\left(x_{0i}^d - x_{0i}^a\right) \right] \\ \quad \geq 2(1-\alpha)x_{0i}^c + (2\alpha-1)x_{0i}^d + s_i^+, i = 1,2\cdots,p \\ \sum_{k=1}^{n} \lambda_k \left[2(1-\alpha)y_{kj}^c + (2\alpha-1)y_{kj}^d \right] \\ \quad + \lambda_0 \left[2(1-\alpha)\left(y_{0j}^b - y_{0j}^c\right) + (2\alpha-1)\left(x_{0j}^a - y_{0i}^d\right) \right] \\ \quad \leq 2(1-\alpha)y_{0j}^b + (2\alpha-1)y_{0j}^a - s_j^-, j = 1,2,\cdots,q \\ \sum_{j=1}^{n} \lambda_k = 1 \\ \lambda_k \geq 0, \quad k = 1,2,\cdots,n \\ s_i^- \geq 0, \quad i = 1,2,\cdots,p \\ s_j^+ \geq 0, \quad j = 1,2\cdots,q \end{cases} \quad (4.31)$$

in which the trapezoidal fuzzy variable $\left(x_{ki}^a, x_{ki}^b, x_{ki}^c, x_{ki}^d\right)$ represents the ith input of DMU$_k$ and $\left(y_{kj}^a, y_{kj}^b, y_{kj}^c, y_{kj}^d\right)$ represents the ith output of DMU$_k$, respectively.

Above two models both are linear programming. Thus, they can be easily solved by many traditional methods.

Table 4.8 Fuzzy optimistic ranking results with $\alpha = 0.8$

DMUs	(v_1, v_2, u_1, u_2)	Efficiency value
DMU_1	(0.0000, 0.0403, 0.0098, 0.0091)	0.6601
DMU_2	(0.0136, 0.0562, 0.0136, 0.0127)	1.0000
DMU_3	(0.0216, 0.0000, 0.0173, 0.0011)	0.7444
DMU_4	(0.2588, 0.0322, 0.0540, 0.3670)	1.0000
DMU_5	(0.0000, 0.0303, 0.0074, 0.0068)	0.9486

Table 4.9 Fuzzy optimistic ranking results with different α

α	DMU_1	DMU_2	DMU_3	DMU_4	DMU_5
0.5	0.592	1	0.723	1	0.911
0.6	0.593	1	0.731	1	0.923
0.7	0.6597	1	0.737	1	0.936
0.8	0.660	1	0.744	1	0.949
0.9	0.665	1	0.75	1	0.962

Table 4.10 Fuzzy maximal chance ranking results with $\alpha = 0.2$

DMUs	$(v_1^*, v_2^*, u_1^*, u_2^*)$	θ^*
DMU_1	(0.044, 0.041, 0.000, 0.166)	0.00
DMU_2	(0.045, 0.041, 0.0000, 0.186)	0.20
DMU_3	(0.4602, 0.0000, 0.0370, 0.6285)	0.00
DMU_4	(0.042, 0.038, 0.000, 0.170)	0.20
DMU_5	(0.017, 0.158, 0.000, 0.070)	0.08

4.4.5 Numerical Examples

This section will give the ranking results by four ranking criteria, respectively. We also use the data in Table 4.3.

Example 4.4 (Fuzzy Optimistic Ranking). Table 4.8 shows the evaluating results by fuzzy optimistic criterion in Sect. 4.4.2 when we set the confidence level $\alpha = 0.8$. According to the ranking criterion, the DMUs can be ranked as follows: DMU_2, DMU_4, DMU_5, DMU_3, DMU_1.

Table 4.9 gives the results of ranking DMUs with different confidence levels. The ranking results are affected by the value α. When $\alpha = 0.9$, the DMUs are ranked: DMU_3, DMU_4, DMU_5, DMU_2, DMU_1. At other confidence levels, the DMUs are ranked as DMU_2, DMU_4, DMU_5, DMU_3, DMU_1.

Example 4.5 (Fuzzy Maximal Chance Ranking). Table 4.10 shows the results of evaluating DMUs by maximal chance ranking criterion in Sect. 4.4.3 with $\alpha = 0.4$. According to the ranking method, the DMUs can be ranked as follows: DMU_2, DMU_4, DMU_5, DMU_3, DMU_1.

4.4 Fuzzy DEA Ranking Criteria

Table 4.11 Fuzzy maximal chance ranking results with different α

Credibility level	DMU$_1$	DMU$_2$	DMU$_3$	DMU$_4$	DMU$_5$
0.50	0.12	0.50	0.22	0.50	0.28
0.40	0.08	0.40	0.20	0.40	0.21
0.30	0.02	0.30	0.19	0.30	0.15
0.20	0	0.20	0.16	0.20	0.08
0.10	0	0.10	0.10	0.10	0.02

Table 4.12 Evaluation results with different α by model (4.30)

α	DMU$_1$	DMU$_2$	DMU$_3$	DMU$_4$	DMU$_5$
0.5	26	0	20	0	0
0.6	22	0	15	0	0
0.7	18	0	10	0	0
0.8	14	0	0	0	0
0.9	3	0	0	0	0

Table 4.13 Evaluation results with different α by model (4.31)

α	DMU$_1$	DMU$_2$	DMU$_3$	DMU$_4$	DMU$_5$
0.5	0	45	0	36	0
0.6	0	48	16	41	0
0.7	0	51	23	46	0
0.8	0	54	29	51	0
0.9	0	57	35	57	0

Table 4.11 gives the results of ranking DMUs with different confidence level α. The greater the value is, the more efficient DMU$_0$ is ranked. The ranking results are varying with different α. When $\alpha = 0.10$, the DMUs are ranked: DMU$_3$, DMU$_4$, DMU$_5$, DMU$_2$, DMU$_1$. At other α, the DMUs are ranked: DMU$_2$, DMU$_4$, DMU$_5$, DMU$_3$, DMU$_1$.

Example 4.6 (Fuzzy Hurwicz Ranking). Table 4.12 gives the evaluation results using model (4.30) with different confidence level α.

Table 4.13 gives the evaluation results using model (4.31) with different confidence level α.

Table 4.14 gives the ranking values with different λ using Hurwicz criterion (4.27) in Sect. 4.4.4. The little the evaluating value is, the better the DMU is ranked. For example, we set $\lambda = 0.5$. Then the DMUs are ranked as follows: DMU$_4$, DMU$_2$, DMU$_5$, DMU$_3$, DMU$_1$. From column 2 ($\lambda = 0$), we know DMU$_1$ and DMU$_5$ are inefficient. From column 12 ($\lambda = 1$), we know DMU$_2$, DMU$_4$, and DMU$_5$ are efficient. DMU$_5$ is both efficient and inefficient. The reason is that DMU$_5$ has the maximal input 1 and input 2 as well as the maximal output 1 and output 2.

Table 4.14 Hurwicz ranking results with different λ

λ	DMU_1	DMU_2	DMU_3	DMU_4	DMU_5
0	0	−51	−23	−46	0
0.1	2	−46	−20	−41	0
0.2	4	−41	−16	−37	0
0.3	5	−36	−13	−31	0
0.4	7	−31	−10	−28	0
0.5	9	−26	−7	−23	0
0.6	11	−20	−3	−18	0
0.7	13	−15	0	−14	0
0.8	14	−10	3	−10	0
0.9	16	−5	7	−5	0
1	18	0	10	0	0

4.5 Fuzzy Congestion

Congestion as one particularly severe form of inefficiency is said to be present when increases in inputs result in output reductions. It was begun in the article by Färe and Svensson [22] in 1980 and then was subsequently given operational implementable form by Färe and Grosskopr [19] in 1983 and Färe et al. [23] in 1985. After then, the pace of research into congestion quickened. It has also awakened interest in the development of alternative approaches as in Cooper et al. [9]. The appearance of a new alternative provides perspective on shortcomings as well as advantages associated with preceding approaches, which is exhibited in the exchanges between Färe and Grosskopf [20], Brokett et al. [1], and Cooper et al. [10, 11].

It has been an under researched topic in the economic theory of production even though it can be of importance when its use is associated with a need for augmenting inputs to serve important objectives besides output maximization. As noted in Cooper et al. [12], for instance, congestion is used in China to deal with the need for providing employment for a large labor force, with some 16,000,000–18,000,000 new entrants each year.

Most of the models on congestion assume that all inputs and outputs data are exactly known, which conflicts with real life in which the data of DMUs cannot be precisely measured. Cooper [16] has challenged how to deal with the congestion in DEA by stochastic method. This section will give some analysis to fuzzy congestion.

4.5.1 Congestion in DEA

As noted in Cooper [13], two approaches to the analysis of congestion are available in DEA literature. One proceeds via radial measure models that use a two-stage approach as described in [23]. Another approach is via the use of additive (nonradial measure) models which also proceed in a two-stage manner [10]. See also the

4.5 Fuzzy Congestion

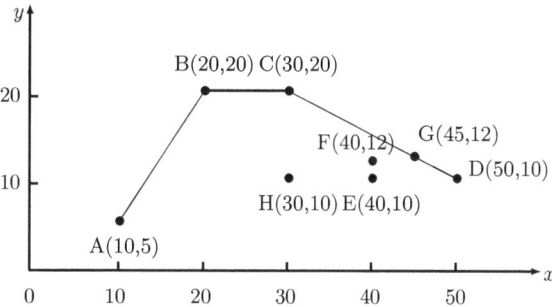

Fig. 4.1 Congestion

exchanges reported in Fare [21] and Cooper [11]. In this paper, we here focus on the development reported in Cooper [12] from which we adopt the following definition:

Congestion: Evidence of congestion is present in the performance of DMU_0 when reductions in one or more inputs are associated with increases in one or more outputs without worsening any other inputs or outputs. More precisely, congestion is evidenced when the attainment of maximal output requires a reduction in one or more of the input amounts used.

We will give some explanation to the definition by Fig. 4.1. On line DC, the maximal output augmentations will always accompany such input reductions, which means the points on this line have congestion. This part differs from the line BA, in accordance with the following definition:

Efficiency (Technical Efficiency): Efficiency is achieved by DMU_0 if and only if it is not possible to improve some of its inputs or outputs without worsening some of its other inputs or outputs.

This property, we may note, is not present in the frontier connecting B and C in Fig. 4.1. This segment of the frontier has the property noted in the following definition:

Inefficiency (Technical Inefficiency): Technical inefficiency is said to be present in the performance of DMU_0 when the evidence shows that it is possible to improve some input or output without worsening some other input or output.

As can be seen in Fig. 4.1, movements from C back to B result in input reductions but not any output increases. The result therefore differs from movements to C from D which identify input decreases that are associated with output increases and, hence, provide the needed evidence of congestion. Thus, as noted in Cooper [12], congestion may be regarded as a particularly severe (separately identifiable) form of technical inefficiency.

Assume that there are n DMUs to be evaluated, whose inputs and outputs are presented as $x_k = (x_{k1}, x_{k2}, \cdots, x_{kp})$ and $y_k = (y_{k1}, y_{k2}, \cdots, y_{kq})$, $k = 1, 2, \cdots, n$. Following Cooper [10], we begin with the additive model [4]:

$$\begin{cases} \max \sum_{i=1}^{p} s_i^- + \sum_{j=1}^{q} s_j^+ \\ \text{subject to:} \\ \quad \sum_{k=1}^{n} \lambda_k x_{ki} = x_{0i} - s_i^-, \quad i = 1, 2, \cdots, p \\ \quad \sum_{k=1}^{n} \lambda_k y_{kj} = y_{0j} + s_j^+, \quad j = 1, 2, \cdots, q \\ \quad \sum_{k=1}^{n} \lambda_k = 1 \\ \quad \lambda_k \geq 0, \quad k = 1, 2, \cdots, n \\ \quad s_i^- \geq 0, \quad i = 1, 2, \cdots, p \\ \quad s_j^+ \geq 0, \quad j = 1, 2, \cdots, q. \end{cases} \quad (4.32)$$

As noted above, congestion is a particularly severe form of technical inefficiency. In order to identify whether congestion might be present and, if so, in what amounts, the following procedure is utilized by Cooper et al. [9]. Inefficiency is a necessary condition for the presence of congestion. Therefore, first use (4.32) to identify whether DMU_0 is inefficient. If found to be inefficient, utilize the following new model:

$$\begin{cases} \max \sum_{i=1}^{m} \delta_i^- \\ \text{subject to:} \\ \quad x_{0i} - s_i^{-*} = \sum_{k=1}^{n} x_{ki} \lambda_k - \delta_i^-, \quad i = 1, 2 \cdots, p \\ \quad y_{0j} + s_j^{+*} = \sum_{k=1}^{n} y_{kj} \lambda_k, \quad j = 1, 2, \cdots, q \\ \quad \sum_{k=1}^{n} \lambda_k = 1 \\ \quad s_i^{-*} \geq \delta_i^-, \quad i = 1, 2 \cdots, p \\ \quad \lambda_k \geq 0, \quad k = 1, 2, \cdots, n. \end{cases} \quad (4.33)$$

Finally, the congesting amount in the total slack associated with s_i^{-*} is defined

$$s_i^{-c*} = s_i^{-*} - \delta_i^{-*}, \quad i = 1, 2, \cdots, p. \quad (4.34)$$

4.5.2 Fuzzy Congestion

Similar to Sect. 4.5.1, firstly use (4.32) to identify whether DMU_0 is inefficient. If found to be inefficient, utilize the following new fuzzy model:

4.5 Fuzzy Congestion

$$\begin{cases} \max \sum_{i=1}^{m} \delta_i^- \\ \text{subject to:} \\ \quad \text{Cr}\left\{\sum_{k=1}^{n} \tilde{x}_{ki}\lambda_k - \delta_i^- \geq \tilde{x}_{0i} - s_i^{-*}\right\} \geq \alpha, i = 1, 2, \cdots, p \\ \quad \text{Cr}\left\{\sum_{k=1}^{n} \tilde{y}_{kj}\lambda_k \geq \tilde{y}_{0j} + s_j^{+*}\right\} \geq \alpha, \quad j = 1, 2 \cdots, q \\ \quad \sum_{k=1}^{n} \lambda_k = 1 \\ \quad \lambda_k \geq 0, \quad k = 1, 2, \cdots, n \\ \quad s_i^{-*} \geq \delta_i^-, \quad i = 1, 2 \cdots, p. \end{cases} \quad (4.35)$$

Finally, the congesting amount in the total slack associated with s_i^{-*} is defined

$$s_i^{-c*} = s_i^{-*} - \delta_i^{-*}, \ i = 1, 2, \cdots, p. \quad (4.36)$$

Therefore, we have the following combined theorem on α-inefficient and congestion:

Theorem 4.1. *At an optimum of (4.2), (4.35), and (4.36), we have the following:*

(a) *If there exists at least one $s_i^{-*} > 0$, $s_j^{+*} > 0$, $1 \leq i \leq p$, $1 \leq j \leq q$, then DMU_0 is α-inefficient.*
(b) *If there exists at least one $s_i^{-c*} > 0$, $1 \leq i \leq p$, then DMU_0 is α-inefficient and congestion is present.*
(c) *If $s^{+*} = 0$ and $s^{-c*} = 0$, then DMU_0 is on a segment of the fuzzy frontier.*

When the inputs and outputs are trapezoidal fuzzy variables, following Liu's book [37], model (4.35) becomes the following linear programming:

$$\begin{cases} \max \sum_{i=1}^{m} \delta_i^- \\ \text{subject to:} \\ \quad \sum_{k=1}^{n} \lambda_k \left[2(1-\alpha)x_{ki}^c + (2\alpha-1)x_{ki}^d\right] \\ \quad +\lambda_0 \left[2(1-\alpha)\left(x_{0i}^b - x_{0i}^c\right) + (2\alpha-1)\left(x_{0i}^a - x_{0i}^d\right)\right] - \delta_i^- \\ \quad \leq 2(1-\alpha)x_{0i}^b + (2\alpha-1)x_{0i}^a - s_i^{-*}, i = 1, 2 \cdots, p \\ \quad \sum_{k=1}^{n} \lambda_k \left[2(1-\alpha)y_{kj}^b + (2\alpha-1)y_{kj}^a\right] \\ \quad +\lambda_0 \left[2(1-\alpha)\left(y_{0j}^c - y_{0j}^b\right) + (2\alpha-1)\left(y_{0j}^d - y_{0j}^a\right)\right] \\ \quad \geq 2(1-\alpha)y_{0j}^c + (2\alpha-1)y_{0j}^d + s_j^{+*}, j = 1, 2, \cdots, q \\ \quad \sum_{k=1}^{n} \lambda_k = 1 \\ \quad \lambda_k \geq 0, \quad k = 1, 2, \cdots, n \\ \quad s_i^{-*} \geq \delta_i^-, \quad i = 1, 2 \cdots, p \end{cases} \quad (4.37)$$

Table 4.15 DMUs with two fuzzy inputs and two fuzzy outputs

DMU$_i$	Input 1	Input 2
A	T(10,1)	T(5,0)
B	T(20,2)	T(20,2)
C	T(30,3)	T(20,2)
D	T(50,5)	T(10,1)
E	T(40,4)	T(10,1)
F	T(40,4)	T(12,1)
G	T(45,4)	T(12,1)
H	T(30,3)	T(10,1)

Table 4.16 Results of evaluating all the DMUs

DMU$_i$	s_j^{+*}	s_i^{-*}	δ_i^{-*}	Congestion
A	0	0	0	0
B	0	0	0	0
C	0	3.5	3.5	0
D	7.8	27.2	3.6	23.6
E	8.2	16.4	2.2	14.2
F	6.2	16.4	8.2	8.2
G	6.2	21.4	2.2	19.2
H	8.2	7	3.4	3.6

in which the trapezoidal fuzzy variable $\left(x_{ki}^a, x_{ki}^b, x_{ki}^c, x_{ki}^d\right)$ represents the ith input of DMU$_k$ and $\left(y_{kj}^a, y_{kj}^b, y_{kj}^c, y_{kj}^d\right)$ represents the ith output of DMU$_k$, respectively.

4.5.3 A Numerical Example

In order to illustrate fuzzy congestion, let us reconsider Fig. 4.1. Assume that all the points have a spread to the center, thus getting some triangular fuzzy variables showed in Table 4.15. There are two fuzzy inputs and two fuzzy outputs. The membership functions are denoted by (c, s) where c is the center and s is the spread.

Table 4.16 shows the results of evaluating all the DMUs when $\alpha = 0.2$. In stage 1, C, D, E, F, G, and H show slacks as indicated by values of ϕ^*, s_r^{+*}, and s_i^{-*}, so they are all inefficient. This fact suggests that they may be associated with congestion. A and B both are efficient. In stage 2, it can be assumed that D, E, F, G, and H have congestions. Since C has slacks, congestion does not exist. The results of evaluating is consistent with Fig. 4.1.

Table 4.17 shows the congestions of all the DMUs with different α. From the results, we can see that the amount of congestion of each DMU is varying when setting α with different values. Even there is no congestion, e.g., H when $\alpha = 0.5$. These phenomena indicate that fuzzy DEA models are more complex than the normal DEA because of the inherent fuzziness contained in inputs and outputs.

4.6 Hybrid Intelligent Algorithm

Table 4.17 Congestion of evaluating DMUs with different α

DMU_i	0.5	0.4	0.3	0.2	0.1
A	0	0	0	0	0
B	0	0	0	0	0
C	0	0	0	0	0
D	1.24	2.16	2.32	2.36	2.44
E	1.00	1.02	1.28	1.42	1.52
F	0.34	0.50	0.55	0.82	0.94
G	1.50	1.64	1.78	1.92	2.06
H	0	0.12	0.24	0.36	0.48

4.6 Hybrid Intelligent Algorithm

4.6.1 Fuzzy Simulations

In the process of solving the above models, a key point is how to estimate the uncertain functions which are defined as functions with uncertain parameters. Due to the complexity, we design some fuzzy simulations to estimate the uncertain function. We write $f(\boldsymbol{u}, \boldsymbol{v}, \boldsymbol{\xi}) = \dfrac{\boldsymbol{v}^T \tilde{\boldsymbol{y}}_0}{\boldsymbol{u}^T \tilde{\boldsymbol{x}}_0}$, in which $\boldsymbol{\xi}$ is the fuzzy vector.

The first fuzzy function is

$$U_1 : (\boldsymbol{u}, \boldsymbol{v}) \to \max \left\{ \overline{f} \mid \mathrm{Cr} \left\{ f(\boldsymbol{u}, \boldsymbol{v}, \boldsymbol{\xi}) \leq \overline{f} \right\} \leq \alpha \right\}. \tag{4.38}$$

In order to compute the fuzzy function (4.38), we randomly generate θ_k from Θ, write $v(k) = (2\mathrm{Cr}\{\theta_k\}) \wedge 1$, and produce $\boldsymbol{\xi}(\theta_k)$, $k = 1, 2, \cdots, N$, respectively. Equivalently, we randomly generate $\boldsymbol{\xi}(\theta_k)$ and write $v_k = \mu(\boldsymbol{\xi}(\theta_k))$ for $k = 1, 2, \cdots, N$, where μ is the membership function of $\boldsymbol{\xi}$. For any number r, we set

$$L(r) = \frac{1}{2} \left(\max_{1 \leq k \leq N} \{ v_k | f(\boldsymbol{u}, \boldsymbol{v}, \boldsymbol{\xi}(\theta_k)) \leq r \} \right.$$
$$\left. + \min_{1 \leq k \leq N} \{ 1 - v_k | f(\boldsymbol{u}, \boldsymbol{v}, \boldsymbol{\xi}(\theta_k)) > r \} \right). \tag{4.39}$$

It follows from monotonicity that we may employ bisection search to find the maximal value r such that $L(r) \leq \alpha$. This value is an estimation of fuzzy function (4.38). We summarize this process as follows:

Step 1. Randomly generate θ_k from Θ, write $v(k) = (2\mathrm{Cr}\{\theta_k\}) \wedge 1$, and produce $\boldsymbol{\xi}(\theta_k)$, $k = 1, 2, \cdots, N$, respectively. Equivalently, we randomly generate $\boldsymbol{\xi}(\theta_k)$ and write $v_k = \mu(\boldsymbol{\xi}(\theta_k))$ for $k = 1, 2, \cdots, N$, where μ is the membership function of $\boldsymbol{\xi}$.
Step 2. Find the maximal value r such that $L(r) \leq \alpha$ holds.
Step 3. Return r.

The second fuzzy function is

$$U_2 : (\boldsymbol{u}, \boldsymbol{v}) \to \mathrm{Cr}\{f(\boldsymbol{u}, \boldsymbol{v}, \boldsymbol{\xi}) \geq 1\}. \tag{4.40}$$

We randomly generate θ_k from Θ, write $\nu(k) = (2\mathrm{Cr}\{\theta_k\}) \wedge 1$, and produce $\boldsymbol{\xi}(\theta_k)$, $k = 1, 2, \cdots, N$, respectively. Equivalently, we randomly generate $\boldsymbol{\xi}(\theta_k)$ and write $\nu_k = \mu(\boldsymbol{\xi}(\theta_k))$ for $k = 1, 2, \cdots, N$, where μ is the membership function of $\boldsymbol{\xi}$. Then U can be estimated by the formula

$$\frac{1}{2}\left(\max_{1\leq k\leq N}\{\nu_k | f(\boldsymbol{u}, \boldsymbol{v}, \boldsymbol{\xi}(\theta_k)) \geq 1\} + \min_{1\leq k\leq N}\{1 - \nu_k | f(\boldsymbol{u}, \boldsymbol{v}, \boldsymbol{\xi}(\theta_k)) < 1\}\right). \tag{4.41}$$

We summarize this process as follows:

Step 1. Randomly generate θ_k from Θ, write $\nu(k) = (2\mathrm{Cr}\{\theta_k\}) \wedge 1$, and produce $\boldsymbol{\xi}(\theta_k)$, $k = 1, 2, \cdots, N$, respectively. Equivalently, we randomly generate $\boldsymbol{\xi}(\theta_k)$ and write $\nu_k = \mu(\boldsymbol{\xi}(\theta_k))$ for $k = 1, 2, \cdots, N$, where μ is the membership function of $\boldsymbol{\xi}$.

Step 2. Return U via the estimation formula.

4.6.2 Genetic Algorithm

Most of the fuzzy DEA models in the proceeding sections have no analytical expressions, then we can try to obtain suboptimal solution by using the heuristic algorithms. Genetic algorithm is a stochastic search method for optimization problems based on the mechanics of natural selection and natural genetics, i.e., survival of the fittest, which has been well documented in the literatures, such as Holland [25] and Koza [30, 31]. In the past decades, genetic algorithm has obtained considerable success in providing satisfactory solutions to many complex optimization problems and received more and more attentions. Since genetic algorithm has solved many uncertain programming models successfully (Liu [35]), here we use it to compute the fuzzy DEA model. The details about genetic algorithm can consult Liu [35].

Chromosome Representation: The first point for genetic algorithm is how to construct a one-to-one mapping between the solution space and the chromosome space such that the following operations such as crossover and mutation can be simplified. The mapping from the solution space to the chromosome space is called encoding, and the mapping from the chromosome space to the solution space is called decoding.

We use a nonnegative vector $\boldsymbol{x} = (v_1, v_2, \cdots, v_q, u_1, u_2, \cdots, u_p)$ to express a decision, in which v_i is the ith coefficient of output and u_j is the jth coefficient of input, $i = 1, 2, \cdots, q$, $j = 1, 2, \cdots, p$.

4.6 Hybrid Intelligent Algorithm

Initialization Process: We randomly generate $x = (v_1, v_2, \cdots, v_q, u_1, u_2, \cdots, u_p)$, in which v_i and u_j are nonnegative numbers, $i = 1, 2, \cdots, q$, $j = 1, 2, \cdots, p$. The feasibility of x can be verified by the fuzzy simulation. If it is feasible, then it will be accepted as a chromosome. If not, then we regenerate a point randomly until a feasible one is obtained. We can make *pop_size* initial feasible chromosomes $x_1, x_2, \cdots, x_{pop\_size}$ by repeating the above process *pop_size* times.

Evaluation Function: Evaluation function, denoted by $Eval(x)$, is to assign a probability of reproduction to each chromosome x. That is, the chromosomes with higher fitness will have more chance to produce offspring.

Firstly we compute the objective values by fuzzy simulations for chromosomes $x_1, x_2, \cdots, x_{pop\_size}$. According to these values, we can give an order relationship among them such that the *pop_size* chromosomes can be rearranged from good to bad. Rewrite them as $x'_1, x'_2, \cdots, x'_{pop\_size}$.

Let a parameter $a \in (0, 1)$ in the genetic system be given. We can define the rank-based evaluation function as follows:

$$Eval(x'_i) = a(1-a)^{i-1}, \quad i = 1, 2, \cdots, pop\_size.$$

Selection Process: During each successive generation, a proportion of the existing population is selected to breed a new generation. The selection process is based on spinning the roulette wheel *pop_size* times. Each time we select a single chromosome for a new population. The roulette wheel is a fitness-proportional selection, where fitter chromosomes (as measured by the objective value) are typically more likely to be selected. The process is always stated as follows:

Step 1. Calculate the cumulative probability q_i for each chromosome x_i,

$$\begin{cases} q_0 = 0, \\ q_i = \sum_{j=1}^{i} Eval(x_j), \quad i = 1, 2, \cdots, pop\_size. \end{cases}$$

Step 2. Generate a random number s in $(0, q_{pop\_size})$.
Step 3. Select the chromosome x_i such that $q_{i-1} < s \leq q_i$.
Step 4. Repeat the second and third steps *pop_size* times and obtain *pop_size* chromosomes.

Crossover Operation: Crossover is one of the mainly used operations for generating a second population. We define a parameter P_c as the probability of crossover. Generating a random number r from the interval $[0, 1]$, the chromosome x_i is selected if $r < P_c$. We denote the selected parents by $x'_1, x'_2, \cdots$. The children of x'_1 and x'_2 are

$$x''_1 = \left(x_1^{(1)} \times u + x_1^{(2)} \times (1-u), x_2^{(1)} \times u + x_2^{(2)} \times (1-u), \cdots, x_n^{(1)} \times u + x_n^{(2)} \times (1-u)\right)$$

and

$$x_2'' = \left(x_1^{(2)} \times u + x_1^{(1)} \times (1-u), x_2^{(2)} \times u + x_2^{(1)} \times (1-u), \cdots, x_n^{(2)} \times u + x_n^{(1)} \times (1-u)\right),$$

in which $u \in (0, 1)$.

We check the feasibility for each child before accepting it. If both children are feasible, then we replace the parents with them. If not, we keep the feasible one if it exists and then redo the crossover operator until two feasible children are obtained or a number of cycles are finished.

Mutation Operation: Mutation is another operation for updating the chromosomes. We define a parameter P_m as the probability of mutation. Generating a random number r from the interval $[0, 1]$, the chromosome x_i is selected if $r < P_m$. Let W be an appropriate large positive number. The child of x_1 is

$$x_1' = (x_1 + d_1 \times w, x_2 + d_2 \times w, \cdots, x_n + d_n \times w)$$

in which $d_i \in [-1, 1]$, $w \in [0, W]$. If x' is not feasible, then we set W as a random number between 0 and W until it is feasible.

4.6.3 Hybrid Intelligent Algorithm

In order to solve the fuzzy DEA model, we integrate the fuzzy simulation and genetic algorithm to produce a hybrid intelligent algorithm. We describe the algorithm as the following procedure:

Step 1. Initialize *pop_size* chromosomes $V_k = (u^k, v^k)$, $k = 1, 2, \ldots, pop\_size$.
Step 2. Calculate the objective values U^k for all chromosomes V_k, $k = 1, 2, \ldots, pop\_size$ by fuzzy simulations, respectively.
Step 3. Compute the fitness of all chromosomes V_k, $k = 1, 2, \ldots, pop\_size$.
Step 4. Select the chromosomes for a new population.
Step 5. Renew the chromosomes V_k, $k = 1, 2, \ldots, pop\_size$ by crossover and mutation operations.
Step 6. Repeat the second to the fifth steps for a given number of cycles.
Step 7. Return the best value.

References

1. Brockett PL, Cooper WW, Hong CS, Wang YY (1998) Inefficiency and congestion in Chinese production before and after the 1978 economic reforms. Socio-Econ Plan Sci 32:1–20
2. Charnes A, Cooper W (1961) Management models and industrial applications of linear programming. Wiley, New York

3. Charnes A, Neralic L (1990) Sensitivity analysis of the additive model in data envelopment analysis. Eur J Oper Res 48:332–341
4. Charnes A, Cooper WW, Golany B, Seiford L, Stutz J (1985) Foundations of data envelopment analysis for Pareto-Koopmans efficient empirical production functions. J Econom 30:91–107
5. Charnes A, Cooper WW, Lewin AY, Morey RC, Rousseau J (1985) Sensitivity and stability analysis in DEA. Ann Oper Res 2:139–156
6. Charnes A, Haag S, Jaska P, Semple J (1992) Sensitivity of efficiency classifications in the additive model of data envelopment analysis. Int J Syst Sci 23:789–798
7. Charnes A, Rousseau J, Semple J (1996) Sensitivity and stability of efficiency classifications in data envelopment analysis. J Product Anal 7:5–18
8. Chvatal V (1983) Linear programming. W. H. Freeman, New York
9. Cooper WW, Thompson RG, Thrall RM (1996) Introduction: extensions and new developments in DEA. Ann Oper Res 66:3–46
10. Cooper WW, Seiford LM, Zhu J (2000) A unified additive model approach for evaluating efficiency and congestion. Socio-Econ Plan Sci 34:1–26
11. Cooper WW, Seiford LM, Zhu J (2001) Slacks and congestion: response to a comment by R. Färe and S. Grosskopf. Socio-Econ Plan Sci 35:205–215
12. Cooper WW, Deng H, Gu B, Li S, Thrall RM (2001) Using DEA to improve the management of congestion in Chinese industries (1981–1997). Socio-Econ Plan Sci 35:1–16
13. Cooper WW, Gu B, Li S (2001) Comparisons and evaluations of alternative approaches to evaluating congestion in DEA. Eur J Oper Res 132:2–74
14. Cooper WW, Park KS, Yu G (2001) An illustrative application of IDEA (imprecise data envelopment analysis) to a Korean mobile telecommunication company. Oper Res 49:807–820
15. Cooper WW, Park KS, Yu G (2001) IDEA (imprecise data envelopment analysis) with CMDs (column maximum decision making units). J Oper Res Soc 52:176–181
16. Cooper WW, Deng H, Huang Z, Li SX (2004) Chance constrained programming approaches to congestion in stochastic data envelopment analysis. Eur J Oper Res 155:487–501
17. Entani T, Maeda Y, Tanaka H (2002) Dual models of interval DEA and its extension to interval data. Eur J Oper Res 136:32–45
18. Fang SC, Puthenpura S (1993) Linear optimization and extensions: theory and algorithms. Prentice-Hall, Englewood Cliffs
19. Färe R, Grosskopf S (1983) Measuring congestion in production. Zeitschrift für Nationalökonomie 43:257–271
20. Färe R, Grosskopf S (1998) Congestion: a note. Socio-Econ Plan Sci 32:21–23
21. Färe R, Grosskopf S (2001) When can slacks be used to identify congestion? An answer to Cooper, W. W. Seiford, L, and Zhu, J. Socio-Econ Plan Sci 35:1–10
22. Färe R, Svensson L (1980) Congestion of factors of production. Econometrica 48:1743–1753
23. Färe R, Grosskopf S, Lovell CAK (1985) The measurement of efficiency of production. Kluwer-Nijhoff Publishing, Boston
24. Guo P, Tanaka H (2001) Fuzzy DEA: a perceptual evaluation method. Fuzzy Sets Syst 119:149–160
25. Holland JH (1975) Adaptation in natural and artificial systems. University of Michigan Press, Ann Arbor
26. Hurwicz L (1951) Optimality criteria for decision making under ignorance. Cowles Commission discussion paper, Chicago, 370
27. Hurwicz L (1951) Some specification problems and application to econometric models (abstract). Econometrica 19:343–344
28. Jahanshahloo GR, Afzalinejad M (2006) A ranking method based on a full-inefficient frontier. Appl Math Model 30:248–260
29. Kao C, Liu ST (2000) Fuzzy efficiency measures in data envelopment analysis. Fuzzy Sets Syst 119:149–160
30. Koza JR (1992) Genetic programming. MIT, Cambridge
31. Koza JR (1994) Genetic programming II. MIT, Cambridge

32. Lertworasirikul S, Fang SC, Joines JA, Nuttle HLW (2003) Fuzzy data envelopment analysis (DEA): a possibility approach. Fuzzy Sets Syst 139:379–394
33. Liu B (1997) Dependent-chance programming: a class of stochastic programming. Comput Math Appl 34(12):89–104
34. Liu B (1999) Dependent-chance programming with fuzzy decisions. IEEE Trans Fuzzy Syst 7:354–360
35. Liu B (2002) Theory and practice of uncertain programming. Physica-Verlag, Heidelberg
36. Liu B (2002) Random fuzzy dependent-chance programming and its hybrid intelligent algorithm. Inf Sci 141(3–4):259–271
37. Liu B (2004) Uncertainty theory. Springer, Berlin
38. Liu B, Liu YK (2002) Expected value of fuzzy variable and fuzzy expected value models. IEEE Trans Fuzzy Syst 10(4):445–450
39. Seiford LM (1994) A DEA bibliography (1978–1992). In: Charnes A, Cooper WW, Lewin A, Seiford L (eds) Data envelopment analysis: theory, methodology and applications. Kluwer Academic, Boston
40. Seiford LM, Zhu J (1998) Stability regions for maintaining efficiency in data envelopment analysis. Eur J Oper Res 108(1):127–139
41. Seiford LM, Zhu J (1999) Infeasibility of super efficiency data envelopment analysis models. INFOR 37(2):174–187
42. Wen ML, Li HS (2009) Fuzzy data envelopment analysis (DEA): model and ranking method. J Comput Appl Math 223:872–878
43. Wen ML, You CL, Kang R (2010) A new ranking method to fuzzy data envelopment analysis. Comput Math Appl 59:3398–3404
44. Wen ML, Zhou D, Lv C (2011) A duzzy data envelopment analysis (DEA) model with credibility measure. Inf Int Interdiscip J 14(6):1947–1958
45. Wen ML, Qin ZF, Kang R (2011) Sensitivity and stability analysis in fuzzy data envelopment analysis (DEA). Fuzzy Optim Decis Mak 10(1):1–10
46. Zadeh LA (1978) Fuzzy sets as a basis for a theory of possibility. Fuzzy Sets Syst 1:3–28
47. Zhu J (1996) Robustness of the efficient DMUs in data envelopment analysis. Eur J Oper Res 90:451–460

Chapter 5
Uncertain DEA

A lot of surveys showed that human uncertainty does not behave like fuzziness. For example, we say "the input is about 10." Generally, we employ fuzzy variable to describe the concept of "about 10"; then there exists a membership function, such as a triangular one (9, 10, 11). Based on this membership function, we can obtain that the lifetime is "exactly 10" with possibility measure 1. On the other hand, the opposite event of "not exactly 10" has the same possibility measure. The conclusion that "not 10" and "exactly 10" have the same possibility measure is not appropriate. In order to have a better mathematical tool to deal with empirical data, uncertainty theory was established by Liu [10] in 2007 and refined in 2010 [14]. As an extension of uncertainty theory, uncertain process and uncertain differential equations (Liu [11]) and uncertain calculus (Liu [12]) were proposed. Uncertain programming was first proposed by Liu [13] in 2009, aiming to deal with the optimal problems involving uncertain variables. This work was followed by an uncertain multiobjective programming and an uncertain goal programming (Liu and Chen [16]) and an uncertain multilevel programming (Liu and Yao [17]). Since then, uncertainty theory was used to solve a variety of real optimal problems, including finance (Chen and Liu [2], Peng [18], Liu [15]), reliability analysis (Liu [7], Zeng et al. [22]), and graph (Gao [3], Gao and Gao [4]). This chapter will give an introduction to uncertain DEA based on uncertain measure.

5.1 Symbols and Notations

This section will introduce an uncertain DEA model, proposed by Wen et al. [19]. The symbols and notations are given as follows:

DMU_i: the ith DMU, $i = 1, 2, \cdots, n$;
DMU_0: the target DMU;
$\tilde{x}_k = (\tilde{x}_{k1}, \tilde{x}_{k2}, \cdots, \tilde{x}_{kp})$: the uncertain inputs vector of DMU_k, $k = 1, 2, \cdots, n$;

$\Phi_{ki}(x)$: the uncertainty distribution of $\tilde{x}_{ki}$, $k = 1, 2, \cdots, n$, $i = 1, 2, \cdots, p$;

$\Phi_k(x) = \big(\Phi_{k1}(x), \Phi_{k2}(x), \cdots, \Phi_{kp}(x)\big)$: the uncertainty distribution vector of $\tilde{x}_k = \big(\tilde{x}_{k1}, \tilde{x}_{k2}, \cdots, \tilde{x}_{kp}\big)$, $k = 1, 2, \cdots, n$;

$x_0 = \big(x_{01}, x_{02}, \cdots, x_{0p}\big)$: the uncertain inputs vector of the target DMU$_0$;

$\Phi_{0i}(x)$: the uncertainty distribution of $\tilde{x}_{0i}$, $i = 1, 2, \cdots, p$;

$\tilde{y}_k = \big(\tilde{y}_{k1}, \tilde{y}_{k2}, \cdots, \tilde{y}_{kq}\big)$: the uncertain outputs vector of DMU$_k$, $k = 1, 2, \cdots, n$;

$\Psi_{kj}(x)$: the uncertainty distribution of $\tilde{x}_{kj}$, $k = 1, 2, \cdots, n$, $j = 1, 2, \cdots, q$;

$\Psi_k(x) = \big(\Psi_{k1}(x), \Psi_{k2}(x), \cdots, \Psi_{kq}(x)\big)$: the uncertainty distribution vector of $\tilde{y}_k = \big(\tilde{y}_{k1}, \tilde{y}_{k2}, \cdots, \tilde{y}_{kq}\big)$, $k = 1, 2, \cdots, n$;

$y_0 = \big(y_{01}, y_{02}, \cdots, y_{0q}\big)$: the outputs vector of the target DMU$_0$;

$\Psi_{0j}(x)$: the uncertainty distribution of $\tilde{x}_{0j}$, $j = 1, 2, \cdots, q$;

$u \in R^{p \times 1}$: the vector of input weights;

$v \in R^{q \times 1}$: the vector of output weights.

5.2 Uncertain DEA Models

This section will consider the DEA when the inputs and outputs are uncertain variables.

Since the uncertain constraints $\sum_{k=1}^{n} \tilde{x}_{ki} \lambda_k \leq \tilde{x}_{0i} - s_i^-$, $i = 1, 2 \cdots, p$ and $\sum_{k=1}^{n} \tilde{y}_{kj} \lambda_k \geq \tilde{y}_{0j} + s_j^+$, $j = 1, 2, \cdots, q$ do not define a deterministic feasible set, a natural idea is to provide a confidence level α, at which it is desired that the uncertain constraints hold. In other words, the event may not happen within $1 - \alpha$. Thus we have some chance constraints as follows:

$$\mathcal{M}\left\{\sum_{k=1}^{n} \tilde{x}_{ki} \lambda_k \leq \tilde{x}_{0i} - s_i^-\right\} \geq \alpha, \quad i = 1, 2, \cdots, p,$$

$$\mathcal{M}\left\{\sum_{k=1}^{n} \tilde{y}_{kj} \lambda_k \geq \tilde{y}_{0j} + s_j^+\right\} \geq \alpha, \quad j = 1, 2 \cdots, q \qquad (5.1)$$

in which $\mathcal{M}$ is the uncertainty measure introduced in Sect. 1.3.

Similar to the deterministic case, the objective of the uncertain model is to maximize the total slacks in inputs and outputs subject to the constraints (5.1). Then Wen et al. [19] proposed the uncertain DEA model:

5.2 Uncertain DEA Models

$$\begin{cases} \max \sum_{i=1}^{p} s_i^- + \sum_{j=1}^{q} s_j^+ \\ \text{subject to:} \\ \qquad \mathcal{M}\left\{\sum_{k=1}^{n} \tilde{x}_{ki}\lambda_k \leq \tilde{x}_{0i} - s_i^-\right\} \geq \alpha, \quad i = 1, 2, \cdots, p \\ \qquad \mathcal{M}\left\{\sum_{k=1}^{n} \tilde{y}_{kj}\lambda_k \geq \tilde{y}_{0j} + s_j^+\right\} \geq \alpha, \quad j = 1, 2 \cdots, q \\ \qquad \sum_{k=1}^{n} \lambda_k = 1 \\ \qquad \lambda_k \geq 0, \quad k = 1, 2, \cdots, n \\ \qquad s_i^- \geq 0, \quad i = 1, 2 \cdots, p \\ \qquad s_j^+ \geq 0, \quad j = 1, 2, \cdots, q. \end{cases} \quad (5.2)$$

Definition 5.1 (Wen et al. [19]). DMU_0 is α-efficient if s_i^{-*} and s_j^{+*} are zero for $i = 1, 2, \cdots, p$ and $j = 1, 2\cdots, q$, where s_i^{-*} and s_j^{+*} are optimal solutions of (5.2).

This definition aligns more closely with Definition 2.5. However, it differs in that uncertain measure is involved. For instance, as determined by the choice of α, there is a risk that DMU_0 will not be efficient even when the condition of Definition 2 is satisfied.

Since $j = 0$ is one of the DMU_j, we can always get a solution with $\lambda_0 = 1, \lambda_j = 0$ ($j \neq 0$), and all slacks zero. Thus this fuzzy DEA model has feasible solution and the optimal value $s_i^{-*} = s_j^{+*} = 0$ for all i, j.

Theorem 5.1. *The objective value in (5.2) is a decreasing function of α.*

Proof. The result can be easily got followed by Monotonicity Theorem.

Following Theorem 5.1, we can get that the bigger the confidence level α is, the more the efficient DMU's number is.

Theorem 5.2 (Wen et al. [19]). *Assume that $\tilde{x}_{1i}, \tilde{x}_{2i}, \cdots, \tilde{x}_{ni}$ are independent uncertain inputs with uncertainty distribution $\Phi_{1i}, \Phi_{2i}, \cdots, \Phi_{ni}$ for each i, $i = 1, 2, \cdots, p$, and $\tilde{y}_{1i}, \tilde{y}_{2i}, \cdots, \tilde{y}_{ni}$ are independent uncertain outputs with uncertainty distribution $\Psi_{1j}, \Psi_{2j}, \cdots, \Psi_{nj}$ for each j, $j = 1, 2, \cdots, q$. Then*

$$\mathcal{M}\left\{\sum_{k=1}^{n} \tilde{x}_{ki}\lambda_k \leq \tilde{x}_{0i} - s_i^-\right\} \geq \alpha, \quad i = 1, 2, \cdots, p$$

$$\mathcal{M}\left\{\sum_{k=1}^{n} \tilde{y}_{kj}\lambda_k \geq \tilde{y}_{0j} + s_j^+\right\} \geq \alpha, \quad j = 1, 2\cdots, q \quad (5.3)$$

holds if and only if

$$\sum_{k=1, k\neq 0}^{n} \lambda_k \Phi_{ki}^{-1}(\alpha) + \lambda_0 \Phi_{0i}^{-1}(1-\alpha) \leq \Phi_{0i}^{-1}(1-\alpha) - s_i^-, \quad i = 1, 2, \cdots, p,$$

$$\sum_{k=1, k\neq 0}^{n} \lambda_k \Psi_{kj}^{-1}(1-\alpha) + \lambda_0 \Psi_{0j}^{-1}(\alpha) \geq \Psi_{0j}^{-1}(\alpha) + s_j^-, \quad j = 1, 2, \cdots, q. \quad (5.4)$$

Proof. Without loss of generality, let $i = 1$ and $x_0 = x_1$; then we will consider the equation

$$\mathcal{M}\left\{\sum_{k=1}^{n} \tilde{x}_{k1}\lambda_k \leq \tilde{x}_{11} - s_i^-\right\} \geq \alpha. \quad (5.5)$$

Rewrite Eq. (5.5) as

$$\mathcal{M}\left\{\sum_{k=2}^{n} \tilde{x}_{k1}\lambda_k - (1-\lambda_1)\tilde{x}_{11} \leq -s_i^-\right\} \geq \alpha. \quad (5.6)$$

Since $-(1-\lambda_1)\tilde{x}_{11}$ is an uncertain variable which is decreasing with respect to $\tilde{x}_{11}$, its inverse uncertainty distribution is

$$\Upsilon_{11}^{-1}(\alpha) = -(1-\lambda_1)\Phi_{11}^{-1}(1-\alpha), \quad 0 < \alpha < 1.$$

For each $2 \leq k \leq n$, $\tilde{x}_{k1}\lambda_k$ is an uncertain variable whose inverse uncertainty distribution is

$$\Upsilon_{k1}^{-1}(\alpha) = \lambda_k \Phi_{k1}^{-1}(\alpha), \quad 0 < \alpha < 1.$$

It follows from the operational law that the inverse uncertainty distribution of the sum $\sum_{k=2}^{n} \tilde{x}_{k1}\lambda_k - (1-\lambda_1)\tilde{x}_{11}$ is

$$\Upsilon^{-1}(\alpha) = \sum_{k=1}^{n} \Upsilon_{21}^{-1}(\alpha)$$

$$= \sum_{k=2}^{n} \lambda_k \Phi_{k1}^{-1}(\alpha) - (1-\lambda_1)\Phi_{11}^{-1}(1-\alpha), \quad 0 < \alpha < 1.$$

From which we may derive the result immediately for $i = 1$ and $x_0 = x_1$. Similarly, we can get other results.

Following Theorem 5.2, the uncertain DEA model (5.2) can be converted to the crisp model as follows:

5.2 Uncertain DEA Models

$$\begin{cases} \max \quad \sum_{i=1}^{p} s_i^- + \sum_{j=1}^{q} s_j^+ \\ \text{subject to:} \\ \sum_{k=1,k\neq 0}^{n} \lambda_k \Phi_{ki}^{-1}(\alpha) + \lambda_0 \Phi_{0i}^{-1}(1-\alpha) \leq \Phi_{0i}^{-1}(1-\alpha) - s_i^-, \quad i=1,2,\cdots,p \\ \sum_{k=1,k\neq 0}^{n} \lambda_k \Psi_{kj}^{-1}(1-\alpha) + \lambda_0 \Psi_{0j}^{-1}(\alpha) \geq \Psi_{0j}^{-1}(\alpha) + s_j^+, \quad j=1,2,\cdots,q \\ \sum_{k=1}^{n} \lambda_k = 1 \\ \lambda_k \geq 0, \quad k=1,2,\cdots,n \\ s_i^- \geq 0, \quad i=1,2\cdots,p \\ s_j^+ \geq 0, \quad j=1,2,\cdots,q \end{cases} \quad (5.7)$$

which is a linear programming. Thus it can be easily solved by many traditional methods.

Example 5.1. For illustration, let DMU$_1$ be the target DMU; then the uncertain DEA model (5.7) can be written as

$$\begin{cases} \max \quad s_1^- + s_2^- + s_1^+ + s_2^+ \\ \text{subject to:} \\ \sum_{k=2}^{5} \lambda_k \Phi_{k1}^{-1}(\alpha) + \lambda_1 \Phi_{11}^{-1}(1-\alpha) \leq \Phi_{11}^{-1}(1-\alpha) - s_1^- \\ \sum_{k=2}^{5} \lambda_k \Phi_{k2}^{-1}(\alpha) + \lambda_1 \Phi_{12}^{-1}(1-\alpha) \leq \Phi_{12}^{-1}(1-\alpha) - s_2^- \\ \sum_{k=2}^{5} \lambda_k \Psi_{k1}^{-1}(1-\alpha) + \lambda_1 \Psi_{11}^{-1}(\alpha) \geq \Psi_{11}^{-1}(\alpha) + s_1^+ \\ \sum_{k=2}^{5} \lambda_k \Psi_{k2}^{-1}(1-\alpha) + \lambda_1 \Psi_{12}^{-1}(\alpha) \geq \Psi_{12}^{-1}(\alpha) + s_2^+ \\ \sum_{k=1}^{5} \lambda_k = 1 \\ \lambda_k \geq 0, \quad k=1,2,\cdots,5 \\ s_1^- \geq 0 \\ s_2^- \geq 0 \\ s_1^+ \geq 0 \\ s_2^+ \geq 0. \end{cases} \quad (5.8)$$

Table 5.1 Results of evaluating the DMUs with $\alpha = 0.6$

DMUs	$(\lambda_1^*, \lambda_2^*, \lambda_3^*, \lambda_4^*, \lambda_5^*)$	$\sum_{i=1}^{p} s_i^{-*} + \sum_{j=1}^{q} s_j^{+*}$	The result of evaluating
DMU_1	(0,0.25,0,0.75,0)	18.9	Inefficiency
DMU_2	(0,1,0,0,0)	0	Efficiency
DMU_3	(0,0,0,0.78,0.22)	15.4	Inefficiency
DMU_4	(0,0,0,1,0)	0	Efficiency
DMU_5	(0,0,0,0,1)	0	Efficiency

Table 5.2 Results of evaluating the DMUs with different confidence level α

α	DMU_1	DMU_2	DMU_3	DMU_4	DMU_5
0.5	Inefficiency	Efficiency	Inefficiency	Efficiency	Efficiency
0.6	Inefficiency	Efficiency	Inefficiency	Efficiency	Efficiency
0.7	Inefficiency	Efficiency	Inefficiency	Efficiency	Efficiency
0.8	Inefficiency	Efficiency	Efficiency	Efficiency	Efficiency
0.9	Efficiency	Efficiency	Efficiency	Efficiency	Efficiency

Table 5.1 shows the results of evaluating DMUs with confidence level $\alpha = 0.6$. The results can be interpreted in the following way: DMU_1 and DMU_3 are inefficient, whereas DMU_2, DMU_4, and DMU_5 are efficient. Moreover, DMU_3 is more efficient than DMU_1 from the total slacks $\sum_{i=1}^{p} s_i^{-*} + \sum_{j=1}^{q} s_j^{+*}$, since they are both inefficient.

Uncertain efficiencies obtained from the model (5.8) for different confidence levels α are shown in Table 5.2. DMU_1 is inefficient at all confidence levels, whereas DMU_2, DMU_4, and DMU_5 are always efficient at all levels. The result that DMU_1 and DMU_3 are efficient at higher levels and inefficient at lower levels coincides with Theorem 5.1. It can be seen that the number of the efficient DMUs is affected by the confidence level α. The higher the confidence level α is, the more the number of efficient DMUs is. These phenomena indicate that uncertain DEA models are more complex than the traditional DEA models because of the inherent uncertainty contained in inputs and outputs.

5.3 Sensitivity and Stability

This section will give some sensitivity and stability analysis to uncertain DEA models.

Theorem 5.3 (Wen et al. [20]). *If DMU_0 is α-inefficient, then the optimal solution satisfying $\lambda_0^*(\alpha) = 0$.*

5.3 Sensitivity and Stability

Proof. We assume the target DMU_0 is DMU_1. That is, $x_0 = x_1$, $y_0 = y_1$. We should prove $\lambda_1 = 0$. For a fixed α, suppose the optimal solution is $(\lambda^*, s^{-*}, s^{+*})$ and the optimal objective value is $\sum_{i=1}^{p} s_i^{-*} + \sum_{j=1}^{q} s_j^{+*}$. If $\lambda_1^* = 0$, then the theorem has been proved. Otherwise let $\lambda_1 > 0$. Since DMU_1 is inefficient, there exists at least one $s_i^{-*} > 0$ or $s_j^{+*} > 0$, $i = 1, 2, \cdots, p$, $j = 1, 2, \cdots, q$. Without loss of generality, we assume $s_1^{-*} > 0$. If $\lambda_1^* = 1$, then $\mathcal{M}\{\tilde{x}_{11} \leq \tilde{x}_{11} - s_1^{-*}\} = 0$. The contradiction implies that $\lambda_1^* \neq 1$:

$$\mathcal{M}\left\{\sum_{k=1}^{n} \tilde{x}_{ki}\lambda_k^* \leq \tilde{x}_{1i} - s_i^{-*}\right\}$$

$$= \mathcal{M}\left\{\sum_{k=2}^{n} \tilde{x}_{ki}\lambda_k^* \leq (1-\lambda_1^*)\tilde{x}_{1i} - s_i^{-*}\right\}$$

$$= \mathcal{M}\left\{\frac{\sum_{k=2}^{n} \tilde{x}_{ki}\lambda_k^*}{1-\lambda_1^*} \leq \tilde{x}_{1i} - \frac{s_i^{-*}}{(1-\lambda_1^*)}\right\}$$

$$\geq \alpha, \ i = 1, 2, \cdots, p.$$

Similarly we can get

$$\mathcal{M}\left\{\sum_{k=1}^{n} \tilde{y}_{kj}\lambda_k^* \geq \tilde{y}_{1j} + s_j^{+*}\right\}$$

$$= \mathcal{M}\left\{\frac{\sum_{k=2}^{n} \tilde{y}_{kj}\lambda_k^*}{1-\lambda_1^*} \geq \tilde{y}_{1j} + \frac{s_j^{+*}}{(1-\lambda_1^*)}\right\}$$

$$\geq \alpha, \ j = 1, 2, \cdots, q.$$

Since $\dfrac{\sum_{k=2}^{n}\lambda_k^*}{1-\lambda_1^*} = 1$, $\left(0, \dfrac{\lambda_2^*}{\sum_{k=2}^{n}\lambda_k^*}, \dfrac{\lambda_3^*}{\sum_{k=2}^{n}\lambda_k^*}, \cdots, \dfrac{\lambda_n^*}{\sum_{k=2}^{n}\lambda_k^*}\right)$ is a feasible solution.

Then the objective value is $\dfrac{1}{1-\lambda_1^*}\left(\sum_{i=1}^{p} s_i^{-*} + \sum_{j=1}^{q} s_j^{+*}\right) > \sum_{i=1}^{p} s_i^{-*} + \sum_{j=1}^{q} s_j^{+*}$, since $0 < \lambda_1^* < 1$, which leads to a contradiction with the assumption. Thus $\lambda_1^* = 0$. The theorem is proved.

Theorem 5.4 (Wen et al. [20]). *If a DMU with $(\tilde{x}_0, \tilde{y}_0)$ is inefficient after evaluating by model (5.2), the new DMU with $(\hat{x}_0, \hat{y}_0) = (\tilde{x}_0 - s^{-*}, \tilde{y}_0 + s^{+*})$ is α-efficient, in which s^{-*} and s^{+*} are optimal solution of (5.2).*

Proof. The efficiency of $(\hat{x}_0, \hat{y}_0)$ is evaluated by solving the problem below:

$$\begin{cases} \max \quad \sum_{i=1}^{p} s_i^- + \sum_{j=1}^{q} s_j^+ \\ \text{subject to:} \\ \mathcal{M}\left\{ \sum_{k=1,k\neq 0}^{n} \tilde{x}_{ki}\lambda_k + \hat{x}_{0i}\lambda_0 \leq \hat{x}_{0i} - s_i^- \right\} \geq \alpha, \quad i = 1,2\cdots,p \\ \mathcal{M}\left\{ \sum_{k=1,k\neq 0}^{n} \tilde{y}_{kj}\lambda_k + \hat{y}_{0j}\lambda_0 \geq \hat{y}_{0j} + s_j^+ \right\} \geq \alpha, \quad j = 1,2,\cdots,q \quad (5.9) \\ \sum_{k=1}^{n} \lambda_k = 1 \\ \lambda_k \geq 0, \quad k = 1,2,\cdots,n \\ s_i^- \geq 0, \quad i = 1,2,\cdots,p \\ s_j^+ \geq 0, \quad j = 1,2\cdots,q. \end{cases}$$

Let an optimal solution be $\left(\hat{\lambda}, \hat{s}^+, \hat{s}^-\right)$. Suppose the DMU with $(\hat{x}_0, \hat{y}_0)$ is inefficient; then $\lambda_0 = 0$ can be obtained by Theorem 4.3. By inserting the formula $(\hat{x}_0, \hat{y}_0) = (\tilde{x}_0 - s^{-*}, \tilde{y}_0 + s^{+*})$ into constraints, we have

$$\mathcal{M}\left\{ \sum_{k=1,k\neq 0}^{n} \tilde{x}_{ki}\hat{\lambda}_k \leq \tilde{x}_{0i} - \hat{s}_i^- - s_i^{-*} \right\} \geq \alpha, \quad i = 1,2\cdots,p,$$

$$\mathcal{M}\left\{ \sum_{k=1,k\neq 0}^{n} \tilde{y}_{kj}\hat{\lambda}_k \geq \tilde{y}_{0j} + \hat{s}_j^+ + s_j^{+*} \right\} \geq \alpha, \quad j = 1,2,\cdots,q.$$

Now we can also write the constraints as

$$\mathcal{M}\left\{ \sum_{k=1,k\neq 0}^{n} \tilde{x}_{ki}\hat{\lambda}_k \leq \tilde{x}_{0i} - \tilde{s}_i^- \right\} \geq \alpha, \quad i = 1,2\cdots,p,$$

$$\mathcal{M}\left\{ \sum_{k=1,k\neq 0}^{n} \tilde{y}_{kj}\hat{\lambda}_k \geq \tilde{y}_{0j} + \tilde{s}_j^+ \right\} \geq \alpha, \quad j = 1,2,\cdots,q$$

5.3 Sensitivity and Stability

where $\tilde{s}^+ = \hat{s}^+ + s^{+*} \geq 0$ and $\tilde{s}^- = \hat{s}^- + s^{-*} \geq 0$. Furthermore, we have

$$\sum_{i=1}^{p} \tilde{s}_i^- + \sum_{j=1}^{q} \tilde{s}_j^+ = \left(\sum_{i=1}^{p} \hat{s}_i^- + s_i^{-*}\right) + \left(\sum_{j=1}^{q} \hat{s}_j^+ + s_j^{+*}\right) \leq \sum_{i=1}^{p} s_i^{-*} \sum_{j=1}^{q} s_j^{+*}$$

since these constraints are a feasible solution for model (5.2) and $\sum_{i=1}^{p} s_i^{-*} + \sum_{j=1}^{q} s_j^{+*}$ is maximal. It follows that we have $\sum_{i=1}^{p} \hat{s}_i^- + \sum_{j=1}^{q} \hat{s}_j^+ = 0$ which implies that all components $\hat{s}_i^-$ and $\hat{s}_j^+$ are zero. Hence uncertain efficiency is achieved as claimed.

Theorem 5.4 has given the stable region when the DMU is inefficient. But we also want to know the efficient radius of efficient DMUs. For this purpose, the following model is proposed:

$$\begin{cases} \min \ \sum_{i=1}^{p} t_i^+ + \sum_{j=1}^{q} t_j^- \\ \text{subject to:} \\ \mathcal{M}\left\{\sum_{k=1, k \neq 0}^{n} \tilde{x}_{ki} \lambda_k \leq \tilde{x}_{0i} + t_i^+\right\} \geq \alpha, \quad i = 1, 2, \cdots, p \\ \mathcal{M}\left\{\sum_{k=1, k \neq 0}^{n} \tilde{y}_{kj} \lambda_k \geq \tilde{y}_{0j} - t_j^-\right\} \geq \alpha, \quad j = 1, 2, \cdots, q \quad (5.10) \\ \sum_{k=1}^{n} \lambda_k = 1 \\ \lambda_k \geq 0, \quad k = 1, 2, \cdots, n \\ t_i^+ \geq 0, \quad i = 1, 2, \cdots, p \\ t_j^- \geq 0, \quad j = 1, 2, \cdots, q. \end{cases}$$

Theorem 5.5 (Wen et al. [20]). *The α-efficient DMU_0 stays α-efficient if $(\hat{x}_0, \hat{y}_0) = (\tilde{x}_0 + t^{+*}, \tilde{y}_0 - t^{-*})$, where t^{+*} and t^{-*} are optimal solution of (5.10).*

Proof. Consider the following DEA model for evaluating the relative efficiency of the adjusted DMU_0:

$$\begin{cases} \max \quad \sum_{i=1}^{p} s_i^- + \sum_{j=1}^{q} s_j^+ \\ \text{subject to:} \\ \mathcal{M}\left\{\sum_{k=1,k\neq 0}^{n} \tilde{x}_{ki}\lambda_k + \left(\tilde{x}_{0i} + t_i^{+*}\right)\lambda_0 \leq \left(\tilde{x}_{0i} + t_i^{+*}\right) - s_i^-\right\} \geq \alpha, \\ \qquad\qquad\qquad\qquad\qquad\qquad\qquad\qquad i = 1,2,\cdots,p \\ \mathcal{M}\left\{\sum_{k=1,k\neq 0}^{n} \tilde{y}_{kj}\lambda_k + \left(\tilde{y}_{0j} - t_j^{-*}\right)\lambda_0 \geq \left(\tilde{y}_{0j} - t_j^{-*}\right) + s_j^+\right\} \geq \alpha, \\ \qquad\qquad\qquad\qquad\qquad\qquad\qquad\qquad j = 1,2,\cdots,q \\ \sum_{k=1}^{n} \lambda_k = 1 \\ \lambda_k \geq 0, \quad k = 1,2,\cdots,n \\ s_i^- \geq 0, \quad i = 1,2,\cdots,p \\ s_j^+ \geq 0, \quad j = 1,2,\cdots,q. \end{cases} \quad (5.11)$$

For a fixed α, let the optimal solution be $\left(\lambda_j^*, \lambda_0^*, s^{-*}, s^{+*}\right)$ and assume that the DMU$_0$ is inefficient. From Theorem 5.3, we get $\lambda_0^* = 0$. Thus this optimal solution is a feasible solution for (5.10). Hence $t_i^{+*} - s_i^{-*} \geq t_i^{+*}$ and $t_j^{-*} - s_j^{+*} \geq t_j^{-*}$, which means $s_i^{-*} = 0$ and s_j^{+*}, $i = 1,2,\cdots,p$, $j = 1,2,\cdots,q$. This leads to a contradiction with the assumption.

Similarly, the uncertain model (5.10) is equivalent to the following deterministic model:

$$\begin{cases} \min \quad \sum_{i=1}^{p} t_i^+ + \sum_{j=1}^{q} t_j^- \\ \text{subject to:} \\ \sum_{k=1,k\neq 0}^{n} \lambda_k \Phi_{ki}^{-1}(\alpha) \leq \Phi_{0i}^{-1}(1-\alpha) + t_i^+, \quad i = 1,2,\cdots,p \\ \sum_{k=1,k\neq 0}^{n} \lambda_k \Psi_{kj}^{-1}(1-\alpha) \geq \Psi_{0j}^{-1}(\alpha) - t_j^-, \quad j = 1,2,\cdots,q \\ \sum_{k=1}^{n} \lambda_k = 1 \\ \lambda_k \geq 0, \quad k = 1,2,\cdots,n \\ t_i^+ \geq 0, \quad i = 1,2,\cdots,p \\ t_j^- \geq 0, \quad j = 1,2,\cdots,q. \end{cases} \quad (5.12)$$

From above analysis, the ranges of inputs and outputs and radius of stability of DMU$_0$ are identified as follows:

5.3 Sensitivity and Stability

Table 5.3 DMUs with two uncertain inputs and two uncertain outputs

DMU_i	Input 1	Input 2	Output 1	Output 2
1	$\mathcal{Z}(35,40,45)$	$\mathcal{Z}(29,31,33)$	$\mathcal{Z}(24,26,28)$	$\mathcal{Z}(38,41,44)$
2	$\mathcal{Z}(29,29,29)$	$\mathcal{Z}(14,15,16)$	$\mathcal{Z}(22,22,22)$	$\mathcal{Z}(33,35,37)$
3	$\mathcal{Z}(44,49,54)$	$\mathcal{Z}(32,36,40)$	$\mathcal{Z}(27,32,37)$	$\mathcal{Z}(43,51,59)$
4	$\mathcal{Z}(34,41,48)$	$\mathcal{Z}(21,23,25)$	$\mathcal{Z}(25,29,33)$	$\mathcal{Z}(55,57,59)$
5	$\mathcal{Z}(59,65,71)$	$\mathcal{Z}(36,41,46)$	$\mathcal{Z}(44,51,58)$	$\mathcal{Z}(65,74,83)$

Table 5.4 Sensitivity analysis results of inefficient DMUs by model (5.7)

DMUs	s_1^{-*}	s_2^{-*}	s_1^{+*}	s_2^{+*}
DMU_1	0	9.3	0.2	9.4
DMU_3	0.3	7.6	0.00	7.5

Table 5.5 Sensitivity analysis results of efficient DMUs by model (5.12)

DMUs	t_1^{+*}	t_2^{+*}	t_1^{-*}	t_2^{-*}
DMU_2	0	2.3	1.6	0.00
DMU_4	0	0.2	0.00	12.1
DMU_5	0.00	0.00	24.2	19.2

1. If DMU_0 is α-inefficient by solving model (5.7), then DMU_0 stays α-inefficient if $(\hat{x}_0, \hat{y}_0) = (\tilde{x}_0 - s^-, \tilde{y}_0 + s^+)$, in which $s^- = \{(s_1^-, \cdots, s_p^-) | 0 \leq s_i^- < s_i^{-*}, i = 1,2,\cdots,p\}$, $s^+ = \{(s_1^+, \cdots, s_q^+) | 0 \leq s_j^+ < s_j^{+*}, j = 1,2,\cdots,q\}$, where s_i^{-*} and s_j^{+*} are optimal solutions of (5.7).
2. If DMU_0 is α-efficient by solving model (5.7), then we use model (5.12) to account for the efficient radius. DMU_0 stays α-efficient if $(\hat{x}_0, \hat{y}_0) = (\tilde{x}_0 + t^+, \tilde{y}_0 - t^-)$, in which $t^+ = \{(t_1, \cdots, t_p) | 0 \leq t_i \leq t_i^{+*}, i = 1,2,\cdots,p\}$ and $t^- = \{(t_1, \cdots, t_q) | 0 \leq t_j \leq t_j^{-*}, j = 1,2,\cdots,q\}$, where t_i^{+*} and t_j^{-*} are optimal solutions of (5.12).

Example 5.2. His example wants to illustrate the sensitivity and the stability of the uncertain DEA model. We also use the data in Table 5.3. From Table 5.1, we get the following results: DMU_1 and DMU_3 are inefficient, whereas DMU_2, DMU_4, and DMU_5 are efficient.

Table 5.4 reports the sensitivity analysis results for inefficient DMUs by model (5.7). In Table 5.4, the columns 2 and 3 report lower bounds of variation ranges of inputs and the columns 4 and 5 are upper bounds of variation ranges of outputs. For instance, DMU_1 in Table 5.4 stays inefficient when $(\hat{x}_{11}, \hat{x}_{12}, \hat{y}_{11}, \hat{y}_{12}) = (\tilde{x}_{11} - r_{x1}, \tilde{x}_{12}, \tilde{y}_{11}, \tilde{y}_{12} + r_{y2})$, in which $0 \leq r_{x2} < 9.3$, $0 \leq r_{y1} < 0.2$, and $0 \leq r_{y2} < 9.4$.

Table 5.5 reports the sensitivity analysis results for efficient DMUs by model (5.12). In Table 5.5, the columns 2 and 3 report upper bounds of variation ranges of inputs and the columns 4 and 5 are lower bounds of variation ranges of outputs. For instance, DMU_4 in Table 5.4 stays efficient when $(\hat{x}_{41}, \hat{x}_{42}, \hat{y}_{41}, \hat{y}_{42}) = (\tilde{x}_{41}, \tilde{x}_{42} + r_{x2}, \tilde{y}_{41}, \tilde{y}_{42} - r_{y2})$, in which $0 \leq r_{x2} \leq 0.2$ and $0 \leq r_{y2} \leq 12.1$.

The similar interpretation can be stated for other rows in Tables 5.4 and 5.5.

5.4 Uncertain DEA Ranking Criteria

In many situations, inputs and outputs are so volatile and complex that they are difficult to measure in an accurate way. This inspired many researchers to apply probability to DEA. As we know, probability or statistics needs a large amount of historical data. In the vast majority of real cases, the sample size is too small (even no sample) to estimate a probability distribution. Then we have to invite some domain experts to evaluate their degree of belief that each event will occur. This section will give some researches to empirical uncertain DEA using the theory introduced in Sect. 1.3. The new symbols and notations are given as follows:

$\tilde{x}_k = (\tilde{x}_{k1}, \tilde{x}_{k2}, \cdots, \tilde{x}_{kp})$: the fuzzy input vectors of DMU_k, $k = 1, 2, \cdots, n$.
$\tilde{y}_k = (\tilde{y}_{k1}, \tilde{y}_{k2}, \cdots, \tilde{y}_{kq})$: the fuzzy output vectors of DMU_k, $k = 1, 2, \cdots, n$.
$\boldsymbol{\Phi}_k(x) = (\Phi_{k1}(x), \Phi_{k2}(x), \cdots, \Phi_{kp}(x))$: the uncertainty distribution vector of $\tilde{x}_k = (\tilde{x}_{k1}, \tilde{x}_{k2}, \cdots, \tilde{x}_{kp})$, $k = 1, 2, \cdots, n$.
$\boldsymbol{\Psi}_k(x) = (\Psi_{k1}(x), \Psi_{k2}(x), \cdots, \Psi_{kq}(x))$: the uncertainty distribution vector of $\tilde{y}_k = (\tilde{y}_{k1}, \tilde{y}_{k2}, \cdots, \tilde{y}_{kq})$, $k = 1, 2, \cdots, n$.

In the following sections, three types of uncertain DEA fully ranking criteria are to be investigated.

5.4.1 Expected Ranking Criterion

Liu [10,12] proposed the expected value operator of uncertain variable and uncertain expected value model. The essential idea of the uncertain expected DEA model is to optimize the expected value of $\dfrac{v^T \tilde{y}_0}{u^T \tilde{x}_0}$ subject to some chance constraints; then we have the first type of the uncertain DEA model proposed by Wen et al. [21]:

$$\begin{cases} \theta = \max_{u,v} E\left[\dfrac{v^T \tilde{y}_0}{u^T \tilde{x}_0}\right] \\ \text{subject to:} \\ \mathcal{M}\{v^T \tilde{y}_k \leq u^T \tilde{x}_k\} \geq \alpha, \ k = 1, 2, \cdots, n \\ u \geq 0 \\ v \geq 0 \end{cases} \quad (5.13)$$

in which $\alpha \in (0.5, 1]$.

5.4 Uncertain DEA Ranking Criteria

Definition 5.2 (Wen et al. [21]). A vector $(u, v) \geq 0$ is called a feasible solution to the uncertain programming model (5.13) if

$$\mathcal{M}\{v^T \tilde{y}_k \leq u^T \tilde{x}_k\} \geq \alpha \tag{5.14}$$

for $k = 1, 2, \cdots, n$.

Definition 5.3 (Wen et al. [21]). A feasible solution (u^*, v^*) is called an expected optimal solution to the uncertain programming model (5.13) if

$$E\left[\frac{v^{*T}\tilde{y}_0}{u^{*T}\tilde{x}_0}\right] \geq E\left[\frac{v^T\tilde{y}_0}{u^T\tilde{x}_0}\right] \tag{5.15}$$

for any feasible solution (u, v).

Expected Ranking Criterion: The greater the optimal objective value is, the more efficient DMU$_0$ is ranked.

Theorem 5.6 (Wen et al. [21]). Assume that $\tilde{x}_{1i}, \tilde{x}_{2i}, \cdots, \tilde{x}_{ni}$ are independent uncertain inputs with uncertainty distribution $\Phi_{1i}, \Phi_{2i}, \cdots, \Phi_{ni}$ for each i, $i = 1, 2, \cdots, p$, and $\tilde{y}_{1i}, \tilde{y}_{2i}, \cdots, \tilde{y}_{ni}$ are independent uncertain outputs with uncertainty distribution $\Psi_{1j}, \Psi_{2j}, \cdots, \Psi_{nj}$ for each j, $j = 1, 2, \cdots, q$. Then the uncertain programming model (5.13) is equivalent to the following model:

$$\begin{cases} \theta = \max_{u,v} \int_0^1 \dfrac{v^T \Psi_0^{-1}(\alpha)}{u^T \Phi_0^{-1}(1-\alpha)} \mathrm{d}\alpha. \\ \text{subject to:} \\ v^T \Psi_k^{-1}(\alpha) \leq u^T \Phi_k^{-1}(1-\alpha), \ k = 1, 2, \cdots, n \\ u \geq 0 \\ v \geq 0. \end{cases} \tag{5.16}$$

Proof. Since the function $\dfrac{v^T \tilde{y}}{u^T \tilde{x}}$ is strictly increasing with respect to $\tilde{y}$ and strictly decreasing with respect to $\tilde{x}$, it follows from Theorem 1.26 that the inverse uncertainty distribution of $\dfrac{v^T \tilde{y}}{u^T \tilde{x}}$ is $\dfrac{v^T \Psi^{-1}(\alpha)}{u^T \Phi^{-1}(1-\alpha)}$. Thus $\mathcal{M}\{v^T \tilde{y}_k \leq u^T \tilde{x}_k\} \geq \alpha$ holds if and only if $v^T \Psi_k^{-1}(\alpha) \leq u^T \Phi_k^{-1}(1-\alpha)$ for $k = 1, 2, \cdots, n$.

By using Theorem 1.30, we obtain

$$E\left[\frac{v^T \tilde{y}_0}{u^T \tilde{x}_0}\right] = \int_0^1 \frac{v^T \Psi_0^{-1}(\alpha)}{u^T \Phi_0^{-1}(1-\alpha)} \mathrm{d}\alpha. \tag{5.17}$$

The theorem is thus verified.

5.4.2 Optimistic Ranking Criterion

Chance-constrained programming (CCP), which was initialized by Charnes and Cooper [1], offers a powerful means for modeling stochastic decision systems. The essential idea of chance-constrained programming is to optimize some critical value with a given confidence level subject to some chance constraints. Inspired by this idea, Liu [14] extended it to uncertain programming models. Assuming that the decision makers want to maximize the optimistic value of the uncertain objective at given confidence level, we have the second type of DEA model:

$$\begin{cases} \max_{u,v} \overline{f} \\ \text{subject to:} \\ \mathcal{M}\left\{\dfrac{v^T \tilde{y}_0}{u^T \tilde{x}_0} \geq \overline{f}\right\} \geq 1 - \alpha \\ \mathcal{M}\left\{v^T \tilde{y}_k \leq u^T \tilde{x}_k\right\} \geq \alpha, \ k = 1, 2, \cdots, n \\ u \geq 0 \\ v \geq 0 \end{cases} \quad (5.18)$$

in which $\alpha \in (0.5, 1]$.

Definition 5.4 (Wen et al. [21]). A feasible solution (u^*, v^*) is called an optimistic optimal solution to the uncertain programming model (5.18) if

$$\max\left\{\overline{f} \mid \mathcal{M}\left\{\dfrac{v^{*T}\tilde{y}_0}{u^{*T}\tilde{x}_0} \geq \overline{f}\right\} \geq 1 - \alpha\right\} \geq \max\left\{\overline{f} \mid \mathcal{M}\left\{\dfrac{v^T \tilde{y}_0}{u^T \tilde{x}_0} \geq \overline{f}\right\} \geq 1 - \alpha\right\}$$

for any feasible solution (u, v).

Optimistic Ranking Criterion: The greater the optimal objective value is, the more efficient DMU_0 is ranked.

Theorem 5.7 (Wen et al. [21]). *Assume that $\tilde{x}_{1i}, \tilde{x}_{2i}, \cdots, \tilde{x}_{ni}$ are independent uncertain inputs with uncertainty distribution $\Phi_{1i}, \Phi_{2i}, \cdots, \Phi_{ni}$ for each i, $i = 1, 2, \cdots, p$, and $\tilde{y}_{1i}, \tilde{y}_{2i}, \cdots, \tilde{y}_{ni}$ are independent uncertain outputs with uncertainty distribution $\Psi_{1j}, \Psi_{2j}, \cdots, \Psi_{nj}$ for each j, $j = 1, 2, \cdots, q$. Then the uncertain programming model (5.18) is equivalent to the following model:*

$$\begin{cases} \max_{u,v} \dfrac{v^T \Psi_0^{-1}(\alpha)}{u^T \Phi_0^{-1}(1-\alpha)} \\ \text{subject to:} \\ v^T \Psi_k^{-1}(\alpha) \leq u^T \Phi_k^{-1}(1-\alpha), \ k = 1, 2, \cdots, n \\ u \geq 0 \\ v \geq 0. \end{cases} \quad (5.19)$$

Proof. By using Theorem 1.26, the theorem can be easily obtained.

By Theorem 5.7, the model (5.18) can be converted to the following linear programming model:

$$\begin{cases} \max_{u,v} v^T \Psi_0^{-1}(\alpha) \\ \text{subject to:} \\ \quad u^T \Phi_0^{-1}(1-\alpha) = 1 \\ \quad v^T \Psi_k^{-1}(\alpha) \leq u^T \Phi_k^{-1}(1-\alpha), \ k = 1, 2, \cdots, n \\ \quad u \geq 0 \\ \quad v \geq 0. \end{cases} \quad (5.20)$$

5.4.3 Maximal Chance Ranking Criterion

Sometimes the decision maker may want to maximize the chance of satisfying the event $\dfrac{v^T \tilde{y}_0}{u^T \tilde{x}_0} \geq 1$. In order to model this type of decision system, Liu [7–9] provided the dependent-chance programming (DCP). Here we carried out the DCP model into the DEA as follows:

$$\begin{cases} \theta = \max_{u,v} \mathcal{M}\left\{ \dfrac{v^T \tilde{y}_0}{u^T \tilde{x}_0} \geq 1 \right\} \\ \text{subject to:} \\ \quad \mathcal{M}\left\{ v^T \tilde{y}_k \leq u^T \tilde{x}_k \right\} \geq \alpha, \ k = 1, 2, \cdots, n \\ \quad u \geq 0 \\ \quad v \geq 0 \end{cases} \quad (5.21)$$

in which $\alpha \in (0.5, 1]$.

Definition 5.5 (Wen et al. [21]). A feasible solution (u^*, v^*) is called a maximal chance optimal solution to the uncertain programming model (5.21) if

$$\mathcal{M}\left\{ \dfrac{v^{*T} \tilde{y}_0}{u^{*T} \tilde{x}_0} \geq 1 \right\} \geq \mathcal{M}\left\{ \dfrac{v^T \tilde{y}_0}{u^T \tilde{x}_0} \geq 1 \right\} \quad (5.22)$$

for any feasible solution (u, v).

Maximal Chance Ranking Criterion: The greater the optimal objective value is, the more efficient DMU_0 is ranked.

Theorem 5.8 (Wen et al. [21]). *Assume that $\tilde{x}_{1i}, \tilde{x}_{2i}, \cdots, \tilde{x}_{ni}$ are independent uncertain inputs with uncertainty distribution $\Phi_{1i}, \Phi_{2i}, \cdots, \Phi_{ni}$ for each i, $i = 1, 2, \cdots, p$, and $\tilde{y}_{1i}, \tilde{y}_{2i}, \cdots, \tilde{y}_{ni}$ are independent uncertain outputs with uncertainty distribution $\Psi_{1j}, \Psi_{2j}, \cdots, \Psi_{nj}$ for each j, $j = 1, 2, \cdots, q$. Then the uncertain programming model (5.21) is equivalent to the following model:*

$$\begin{cases} \theta = \max\limits_{u,v} \mathcal{M}\left\{\dfrac{v^T \tilde{y}_0}{u^T \tilde{x}_0} \geq 1\right\} \\ \text{subject to:} \\ \quad v^T \Psi_k^{-1}(\alpha) \leq u^T \Phi_k^{-1}(1-\alpha), \ k = 1, 2, \cdots, n \\ \quad u \geq 0 \\ \quad v \geq 0. \end{cases} \quad (5.23)$$

Proof. By using Theorem 1.26, the theorem can be easily obtained.

5.4.4 Hurwicz Ranking Criterion

The fourth ranking method is the Hurwicz ranking criterion. Section 5.2 has given the uncertain DEA model

$$\begin{cases} \theta_1 = \max \ \sum\limits_{i=1}^{p} s_i^- + \sum\limits_{j=1}^{q} s_j^+ \\ \text{subject to:} \\ \quad \mathcal{M}\left\{\sum\limits_{k=1}^{n} \tilde{x}_{ki} \lambda_k \leq \tilde{x}_{0i} - s_i^-\right\} \geq \alpha, \ i = 1, 2, \cdots, p \\ \quad \mathcal{M}\left\{\sum\limits_{k=1}^{n} \tilde{y}_{kj} \lambda_k \geq \tilde{y}_{0j} + s_j^+\right\} \geq \alpha, \ j = 1, 2 \cdots, q \\ \quad \sum\limits_{k=1}^{n} \lambda_k = 1 \\ \quad \lambda_k \geq 0, \ k = 1, 2, \cdots, n \\ \quad s_i^- \geq 0, \ i = 1, 2 \cdots, p \\ \quad s_j^+ \geq 0, \ j = 1, 2, \cdots, q \end{cases} \quad (5.24)$$

which considers the total distances to an efficient frontier. We call it uncertain optimistic model. We can also give an uncertain DEA model, which considers the total distances to an inefficient frontier:

5.4 Uncertain DEA Ranking Criteria

$$\begin{cases} \theta_2 = \max \sum_{i=1}^{p} s_i^- + \sum_{j=1}^{q} s_j^+ \\ \text{subject to:} \\ \mathcal{M}\left\{\sum_{k=1}^{n} \tilde{x}_{ki}\lambda_k \geq \tilde{x}_{0i} + s_i^-\right\} \geq \alpha, \quad i = 1, 2, \cdots, p \\ \mathcal{M}\left\{\sum_{k=1}^{n} \tilde{y}_{kj}\lambda_k \leq \tilde{y}_{0j} - s_j^+\right\} \geq \alpha, \quad j = 1, 2 \cdots, q \\ \sum_{j=1}^{n} \lambda_j = 1 \\ \lambda_j \geq 0, \quad j = 1, 2, \cdots, n \\ s_i^- \geq 0, \quad i = 1, 2, \cdots, p \\ s_j^+ \geq 0, \quad j = 1, 2 \cdots, q. \end{cases} \quad (5.25)$$

Definition 5.6. DMU_0 is α-inefficient if s_i^{-*} and s_j^{+*} are zero for $i = 1, 2, \cdots, p$ and $j = 1, 2 \cdots, q$, where s_i^{-*} and s_j^{+*} are optimal solutions of (5.25).

Since $j = 0$ is one of the DMU_j, we can always get a solution with $\lambda_0 = 1, \lambda_j = 0 \ (j \neq 0)$, and all slacks zero. Thus the uncertain DEA models (5.2) and (5.25) have feasible solution and the optimal value $s_i^{-*} = s_j^{+*} = 0$ for all i, j.

Abovementioned two models are both extreme cases: one is too optimistic and the other is too pessimistic. Thus, we employ the Hurwicz criterion, suggested by Leonid Hurwicz [5, 6] in 1951, which incorporates a measure of both by assigning a certain percentage weight λ to θ_1^* and $1 - \lambda$ to $-\theta_2^*, 0 \leq \lambda \leq 1$:

$$\theta^* = \lambda\theta_1^* + (1-\lambda)(-\theta_2^*) \quad (5.26)$$

which can be rewritten as

$$\theta^* = \lambda\theta_1^* - (1-\lambda)\theta_2^*. \quad (5.27)$$

Ranking Criterion: The greater the value θ^* is, the less efficient DMU_0 is ranked.

In the Hurwicz criterion, the parameter $\lambda \in [0, 1]$, which reflects the degree of the decision maker's optimism, must be determined by the decision maker. Generally speaking, it is difficult to determine the appropriate λ for decision makers, since it varies from person to person. By varying the parameter λ, the Hurwicz criterion becomes various models, e.g., when $\lambda = 1$, the criterion is the traditional DEA model (5.2); when $\lambda = 0$, it degenerates to model (5.25). This fact means that the Hurwicz criterion is fairly flexible.

In some cases, $\theta_1^* = 0$ and $\theta_2^* = 0$, and then the ranking value $\theta^* = 0$. It says DMU$_0$ is both efficient and inefficient. This happens when DMU$_0$ is the best in some inputs and/or outputs while it is the worst in some other inputs and/or outputs. For example, if DMU$_0$ is the only DMU that has the largest value for input 1 and has the least amount of input 2, DMU$_0$ is both efficient and inefficient.

From the properties of the α-optimistic and α-pessimistic values, the uncertain models (5.2) and (5.25) can be converted to the following linear programming models:

$$\begin{cases} \max \sum_{i=1}^{p} s_i^- + \sum_{j=1}^{q} s_j^+ \\ \text{subject to:} \\ \sum_{k=1,k\neq 0}^{n} \lambda_k \Phi_{ki}^{-1}(\alpha) + \lambda_0 \Phi_{0i}^{-1}(1-\alpha) \leq \Phi_{0i}^{-1}(1-\alpha) - s_i^-, \ i = 1,2,\cdots,p \\ \sum_{k=1,k\neq 0}^{n} \lambda_k \Psi_{kj}^{-1}(1-\alpha) + \lambda_0 \Psi_{0j}^{-1}(\alpha) \geq \Psi_{0j}^{-1}(\alpha) + s_j^+, \ j = 1,2,\cdots,q \\ \sum_{k=1}^{n} \lambda_k = 1 \\ \lambda_k \geq 0, \ k = 1,2,\cdots,n \\ s_i^- \geq 0, \ i = 1,2\cdots,p \\ s_j^+ \geq 0, \ j = 1,2,\cdots,q \end{cases} \quad (5.28)$$

and

$$\begin{cases} \max \sum_{i=1}^{p} s_i^- + \sum_{j=1}^{q} s_j^+ \\ \text{subject to:} \\ \sum_{k=1,k\neq 0}^{n} \lambda_k \Phi_{ki}^{-1}(1-\alpha) + \lambda_0 \Phi_{0i}^{-1}(\alpha) \geq \Phi_{0i}^{-1}(\alpha) + s_i^-, \ i = 1,2,\cdots,p \\ \sum_{k=1,k\neq 0}^{n} \lambda_k \Psi_{kj}^{-1}(\alpha) + \lambda_0 \Psi_{0j}^{-1}(1-\alpha) \leq \Psi_{0j}^{-1}(1-\alpha) - s_j^+, \ j = 1,2,\cdots,q \\ \sum_{k=1}^{n} \lambda_k = 1 \\ \lambda_k \geq 0, \ k = 1,2,\cdots,n \\ s_i^- \geq 0, \ i = 1,2\cdots,p \\ s_j^+ \geq 0, \ j = 1,2,\cdots,q. \end{cases} \quad (5.29)$$

Above two models both are linear programming. Thus they can be easily solved by many traditional methods.

5.5 Uncertain Congestion

Different from fuzzy congestion in Sect. 4.5.2, this section will introduce some congestions with uncertain inputs and outputs. After identifying that DMU_0 is inefficient by model (5.2), the following uncertain model is utilized:

$$\begin{cases} \max \sum_{i=1}^{m} \delta_i^- \\ \text{subject to:} \\ \quad \mathcal{M}\left\{ \sum_{k=1}^{n} \tilde{x}_{ki} \lambda_k - \delta_i^- \geq \tilde{x}_{0i} - s_i^{-*} \right\} \geq \alpha, i = 1, 2, \cdots, p \\ \quad \mathcal{M}\left\{ \sum_{k=1}^{n} \tilde{y}_{kj} \lambda_k \geq \tilde{y}_{0j} + s_j^{+*} \right\} \geq \alpha, \quad j = 1, 2 \cdots, q \\ \quad \sum_{k=1}^{n} \lambda_k = 1 \\ \quad \lambda_k \geq 0, \quad k = 1, 2, \cdots, n \\ \quad s_i^{-*} \geq \delta_i^-, \quad i = 1, 2 \cdots, p \end{cases} \quad (5.30)$$

in which s_i^{-*} and s_j^{+*} are optimal solutions of (5.2).

Finally, the congesting amount in the total slack associated with s_i^{-*} is defined by

$$s_i^{-c*} = s_i^{-*} - \delta_i^{-*}, \ i = 1, 2, \cdots, p. \quad (5.31)$$

Therefore, we have the following combined theorem on α-inefficient and congestion:

Theorem 5.9. *At an optimum of (5.2), (5.30), and (5.31), we have the following:*

(a) *If there exists at least one $s_i^{-*} > 0$, $s_j^{+*} > 0$, $1 \leq i \leq p$, $1 \leq j \leq q$, then DMU_0 is α-inefficient.*
(b) *If there exists at least one $s_i^{-c*} > 0$, $1 \leq i \leq p$, then DMU_0 is α-inefficient and congestion is present.*
(c) *If $s^{+*} = 0$ and $s^{-c*} = 0$, then DMU_0 is on a segment of the uncertain frontier.*

Following Theorem 5.2, the uncertain DEA model (5.2) can be converted to the crisp model as follows:

$$\begin{cases} \max \quad \sum_{i=1}^{p} s_i^- + \sum_{j=1}^{q} s_j^+ \\ \text{subject to:} \\ \sum_{k=1, k \neq 0}^{n} \lambda_k \Phi_{ki}^{-1}(1-\alpha) + \lambda_0 \Phi_{0i}^{-1}(\alpha) - \delta_i^- \geq \Phi_{0i}^{-1}(\alpha) - s_i^{-*}, \quad i = 1, 2, \cdots, p \\ \sum_{k=1, k \neq 0}^{n} \lambda_k \Psi_{kj}^{-1}(1-\alpha) + \lambda_0 \Psi_{0j}^{-1}(\alpha) \geq \Psi_{0j}^{-1}(\alpha) + s_j^{+*}, \quad j = 1, 2, \cdots, q \\ \sum_{k=1}^{n} \lambda_k = 1 \\ \lambda_k \geq 0, \quad k = 1, 2, \cdots, n \\ s_i^- \geq 0, \quad i = 1, 2 \cdots, p \\ s_j^+ \geq 0, \quad j = 1, 2, \cdots, q \end{cases} \quad (5.32)$$

which is a linear programming. Thus it can be easily solved by many traditional methods.

References

1. Charnes A, Cooper W (1961) Management models and industrial applications of linear programming. Wiley, New York
2. Chen X, Liu B (2010) Existence and uniqueness theorem for uncertain differential equations. Fuzzy Optim Decis Mak 9(1):69–81
3. Gao XL (2013) Cycle index of uncertain graph. Inf Int Interdiscip J 16(2)(A):131–1138
4. Gao XL, Gao Y (2013) Connectedness index of uncertainty graphs. Int J Uncertain Fuzziness Knowl-Based Syst 21(1):127–137
5. Hurwicz L (1951) Optimality criteria for decision making under ignorance. Cowles Commission discussion paper, Chicago, 370
6. Hurwicz L (1951) Some specification problems and application to econometric models (abstract). Econometrica 19:343–344 (1951)
7. Liu B (1997) Dependent-chance programming: a class of stochastic programming. Comput Math Appl 34(12):89–104
8. Liu B (1999) Dependent-chance programming with fuzzy decisions. IEEE Trans Fuzzy Syst 7:354–360
9. Liu B (2002) Random fuzzy dependent-chance programming and its hybrid intelligent algorithm. Inf Sci 141(3–4):259–271
10. Liu B (2007) Uncertainty theory, 2nd edn. Springer, Berlin
11. Liu B (2008) Fuzzy process, hybrid process and uncertain process. J Uncertain Syst 2(1):3–16
12. Liu B (2009) Some research problems in uncertainty theory. J Uncertain Syst 3(1):3–10
13. Liu B (2009) Theory and practice of uncertain programming, 2nd edn. Springer, Berlin

References

14. Liu B (2010) Uncertainty theory: a branch of mathematics for modeling human uncertainty. Springer, Berlin
15. Liu B (2013) Extreme value theorems of uncertain process with application to insurance risk model. Soft Comput 17(4):549–556
16. Liu B, Chen XW (2013) Uncertain multiobjective programming and uncertain goal programming. Technical report
17. Liu B, Yao K, Uncertain multilevel programming: algorithm and application. http://orsc.edu.cn/online/120114.pdf
18. Peng J, Yao K (2010) A new option pricing model for stocks in uncertainty markets. Int J Oper Res 7(4):213–224
19. Wen ML, Kang V, Data envelopment analysis (DEA) with uncertain inputs and outputs. http://orsc.edu.cn/online/120514.pdf
20. Wen ML, Qin ZF, Kang R, Sensitivity and stability analysis in uncertain data envelopment (DEA). http://orsc.edu.cn/online/120515.pdf
21. Wen ML, Qin ZF, Kang R, Some new ranking criteria in data envelopment analysis under uncertain environment. http://orsc.edu.cn/online/120.pdf
22. Zeng ZG, Wen ML, Kang R (2013) Belief reliability: a new metrics for products' reliability. Fuzzy Optim Decis Mak 12(1):15–27

Chapter 6
Hybrid DEA

In many cases, uncertainty and randomness simultaneously appear in a system. For example, some DMUs have no samples, while others have enough samples to determine probability distributions. In this case, this chapter will give some hybrid DEA models to deal with the hybrid uncertainties. This chapter will employ chance theory to model the hybrid DEA models.

6.1 Symbols and Notations

This section will introduce the symbols and notations, which are crucial for following sections:

DMU_i: the ith DMU, $i = 1, 2, \cdots, n$;
DMU_0: the target DMU;
$\tilde{x}_k = (\tilde{x}_{k1}, \tilde{x}_{k2}, \cdots, \tilde{x}_{kp})$: the uncertain random inputs vector of DMU_k, $k = 1, 2, \cdots, n$;
$\Phi_{ki}(x)$: the uncertainty random distribution of $\tilde{x}_{ki}$, $k = 1, 2, \cdots, n$, $i = 1, 2, \cdots, p$;
$x_0 = (x_{01}, x_{02}, \cdots, x_{0p})$: the uncertain random inputs vector of the target DMU_0;
$\Phi_{0i}(x)$: the uncertainty random distribution of $\tilde{x}_{0i}$, $i = 1, 2, \cdots, p$;
$\tilde{y}_k = (\tilde{y}_{k1}, \tilde{y}_{k2}, \cdots, \tilde{y}_{kq})$: the uncertain random outputs vector of DMU_k, $k = 1, 2, \cdots, n$;
$\Psi_{kj}(x)$: the uncertainty random distribution of $\tilde{x}_{kj}$, $k = 1, 2, \cdots, n$, $j = 1, 2, \cdots, q$;
$y_0 = (y_{01}, y_{02}, \cdots, y_{0q})$: the uncertainty random outputs vector of the target DMU_0;
$\Psi_{0j}(x)$: the uncertainty random distribution of $\tilde{x}_{0j}$, $j = 1, 2, \cdots, q$;

6.2 Hybrid DEA Models

Similar to the deterministic case, the hybrid DEA model can be given as follows:

$$\begin{cases} \max \quad \sum_{i=1}^{p} s_i^- + \sum_{j=1}^{q} s_j^+ \\ \text{subject to:} \\ \quad \text{Ch}\left\{\sum_{k=1}^{n} \tilde{x}_{ki}\lambda_k \leq \tilde{x}_{0i} - s_i^-\right\} \geq \alpha, \quad i = 1, 2, \cdots, p \\ \quad \text{Ch}\left\{\sum_{k=1}^{n} \tilde{y}_{kj}\lambda_k \geq \tilde{y}_{0j} + s_j^+\right\} \geq \alpha, \quad j = 1, 2\cdots, q \\ \quad \sum_{k=1}^{n} \lambda_k = 1 \\ \quad \lambda_k \geq 0, \quad k = 1, 2, \cdots, n \\ \quad s_i^- \geq 0, \quad i = 1, 2\cdots, p \\ \quad s_j^+ \geq 0, \quad j = 1, 2, \cdots, q \end{cases} \quad (6.1)$$

in which Ch is the chance measure introduced in Sect. 1.4.

Definition 6.1. DMU_0 is α-efficient if s_i^{-*} and s_j^{+*} are zero for $i = 1, 2, \cdots, p$ and $j = 1, 2\cdots, q$, where s_i^{-*} and s_j^{+*} are optimal solutions of (6.1).

Special Case 1: When all the inputs and outputs degenerate to random variables, the model (6.1) becomes the following form:

$$\begin{cases} \max \quad \sum_{i=1}^{p} s_i^- + \sum_{j=1}^{q} s_j^+ \\ \text{subject to:} \\ \quad \Pr\left\{\sum_{k=1}^{n} \tilde{x}_{ki}\lambda_k \leq \tilde{x}_{0i} - s_i^-\right\} \geq \alpha, \quad i = 1, 2, \cdots, p \\ \quad \Pr\left\{\sum_{k=1}^{n} \tilde{y}_{kj}\lambda_k \geq \tilde{y}_{0j} + s_j^+\right\} \geq \alpha, \quad j = 1, 2\cdots, q \\ \quad \sum_{k=1}^{n} \lambda_k = 1 \\ \quad \lambda_k \geq 0, \quad k = 1, 2, \cdots, n \\ \quad s_i^- \geq 0, \quad i = 1, 2\cdots, p \\ \quad s_j^+ \geq 0, \quad j = 1, 2, \cdots, q \end{cases} \quad (6.2)$$

in which Pr is the probability measure introduced in Sect. 1.1.

6.2 Hybrid DEA Models

Special Case 2: When all the inputs and outputs degenerate to uncertain variables, the model (6.1) becomes the following form:

$$\begin{cases} \max \quad \sum_{i=1}^{p} s_i^- + \sum_{j=1}^{q} s_j^+ \\ \text{subject to:} \\ \qquad \mathcal{M}\left\{\sum_{k=1}^{n} \tilde{x}_{ki}\lambda_k \leq \tilde{x}_{0i} - s_i^-\right\} \geq \alpha, \quad i = 1, 2, \cdots, p \\ \qquad \mathcal{M}\left\{\sum_{k=1}^{n} \tilde{y}_{kj}\lambda_k \geq \tilde{y}_{0j} + s_j^+\right\} \geq \alpha, \quad j = 1, 2 \cdots, q \\ \qquad \sum_{k=1}^{n} \lambda_k = 1 \\ \qquad \lambda_k \geq 0, \quad k = 1, 2, \cdots, n \\ \qquad s_i^- \geq 0, \quad i = 1, 2 \cdots, p \\ \qquad s_j^+ \geq 0, \quad j = 1, 2, \cdots, q \end{cases} \quad (6.3)$$

which is just the uncertain DEA model (5.2). Since this model can be converted to the linear programming model, it can be easily solved by many traditional methods. The detailed can refer to Sect. 5.2.

Special Case 3: When some inputs (outputs) are uncertain variables and some inputs (outputs) are random variables, we will give some new symbols:

$\tilde{x}_{ki}$: the ith uncertain input of DMU$_k$, $k = 1, 2, \cdots, n$, $i = 1, 2, \cdots, m$;
$\Phi_{ki}(x)$: the uncertain distribution of $\tilde{x}_{ki}$, $k = 1, 2, \cdots, n$, $i = 1, 2, \cdots, m$;
$\hat{x}_{ki}$: the ith random input of DMU$_k$, $k = 1, 2, \cdots, n$, $i = m+1, m+2, \cdots, p$;
$\Psi_{ki}(x)$: the random distribution of $\tilde{x}_{ki}$, $k = 1, 2, \cdots, n$, $i = m+1, m+2, \cdots, p$;
$\boldsymbol{x}_k = (\tilde{x}_{k1}, \tilde{x}_{km}, \hat{x}_{km+1}, \cdots, \hat{x}_{kp})$: the uncertain random inputs vector of DMU$_k$, $k = 1, 2, \cdots, n$;
$\tilde{y}_{kj}$: the jth uncertain output of DMU$_k$, $k = 1, 2, \cdots, n$, $j = 1, 2, \cdots, l$;
$\Upsilon_{ki}(y)$: the uncertain distribution of $\tilde{y}_{kj}$, $k = 1, 2, \cdots, n$, $j = 1, 2, \cdots, l$;
$\hat{y}_{kj}$: the jth random output of DMU$_k$, $k = 1, 2, \cdots, n$, $j = l+1, l+2, \cdots, q$;
$\Xi_{ki}(y)$: the random distribution of $\hat{y}_{ki}$, $k = 1, 2, \cdots, n$, $j = l+1, l+2, \cdots, q$;
$\boldsymbol{y}_k = (\tilde{y}_{k1}, \tilde{y}_{kl}, \hat{y}_{k\ l+1}, \cdots, \hat{y}_{kq})$: the uncertain random outputs vector of DMU$_k$, $k = 1, 2, \cdots, n$;

Then the model (6.1) has the following form:

$$\begin{cases} \max \quad \sum_{i=1}^{p} s_i^- + \sum_{j=1}^{q} s_j^+ \\ \text{subject to:} \\ \quad \mathcal{M}\left\{\sum_{k=1}^{n} \tilde{x}_{ki}\lambda_k \leq \tilde{x}_{0i} - s_i^-\right\} \geq \alpha, i = 1, 2, \cdots, m \\ \quad \Pr\left\{\sum_{k=1}^{n} \hat{x}_{ki}\lambda_k \leq \tilde{x}_{0i} - s_i^-\right\} \geq \alpha, i = m+1, m+2, \cdots, p \\ \quad \mathcal{M}\left\{\sum_{k=1}^{n} \tilde{y}_{kj}\lambda_k \geq \tilde{y}_{0j} + s_j^+\right\} \geq \alpha, j = 1, 2, \cdots, l \\ \quad \Pr\left\{\sum_{k=1}^{n} \hat{y}_{kj}\lambda_k \geq \tilde{y}_{0j} + s_j^+\right\} \geq \alpha, j = l+1, l+2, \cdots, q \\ \quad \sum_{k=1}^{n} \lambda_k = 1 \\ \quad \lambda_k \geq 0, \quad k = 1, 2, \cdots, n \\ \quad s_i^- \geq 0, \quad i = 1, 2 \cdots, p \\ \quad s_j^+ \geq 0, \quad j = 1, 2, \cdots, q \end{cases} \quad (6.4)$$

in which $\mathcal{M}$ is the uncertainty measure introduced in Sect. 1.3 and Pr is the probability measure introduced in Sect. 1.1.

6.3 Hybrid DEA Ranking Criteria

This section will introduce some fully ranking methods in hybrid DEA. Four types of hybrid uncertain DEA fully ranking criteria are to be investigated.

6.3.1 Expected Ranking Criterion

The essential idea of the hybrid expected DEA model is to optimize the expected value of $\dfrac{v^T \tilde{y}_0}{u^T \tilde{x}_0}$ subject to some chance constraints, then we have the first type of the hybrid DEA model:

6.3 Hybrid DEA Ranking Criteria

$$\begin{cases} \theta = \max_{u,v} E\left[\dfrac{v^T \tilde{y}_0}{u^T \tilde{x}_0}\right] \\ \text{subject to:} \\ \quad \text{Ch}\left\{v^T \tilde{y}_k \leq u^T \tilde{x}_k\right\} \geq \alpha, \ k = 1, 2, \cdots, n \\ \quad u \geq 0 \\ \quad v \geq 0 \end{cases} \quad (6.5)$$

in which $\alpha \in (0.5, 1]$.

Definition 6.2. A vector $(u, v) \geq 0$ is called a feasible solution to the hybrid programming model (6.5) if

$$\mathcal{M}\left\{v^T \tilde{y}_k \leq u^T \tilde{x}_k\right\} \geq \alpha \quad (6.6)$$

for $k = 1, 2, \cdots, n$.

Definition 6.3. A feasible solution (u^*, v^*) is called an expected optimal solution to the hybrid programming model (6.5) if

$$E\left[\dfrac{v^{*T} \tilde{y}_0}{u^{*T} \tilde{x}_0}\right] \geq E\left[\dfrac{v^T \tilde{y}_0}{u^T \tilde{x}_0}\right] \quad (6.7)$$

for any feasible solution (u, v).

Expected Ranking Criterion: The greater the optimal objective value is, the more efficient DMU_0 is ranked.

6.3.2 Optimistic Ranking Criterion

Chance-constrained programming (CCP), which was initialized by Charnes and Cooper [1], offers a powerful means for modeling uncertain decision systems. The essential idea of chance-constrained programming is to optimize some critical value with a given confidence level subject to some chance constraints. Assuming that the decision makers want to maximize the optimistic value of the hybrid objective at given confidence level, we have the second type of DEA model:

$$\begin{cases} \max_{u,v} \overline{f} \\ \text{subject to:} \\ \quad \text{Ch}\left\{ \dfrac{v^T \tilde{y}_0}{u^T \tilde{x}_0} \geq \overline{f} \right\} \geq 1 - \alpha \\ \quad \text{Ch}\left\{ v^T \tilde{y}_k \leq u^T \tilde{x}_k \right\} \geq \alpha, \ k = 1, 2, \cdots, n \\ \quad u \geq 0 \\ \quad v \geq 0 \end{cases} \quad (6.8)$$

in which $\alpha \in (0.5, 1]$.

Definition 6.4. A feasible solution (u^*, v^*) is called an optimistic optimal solution to the hybrid programming model (6.8) if

$$\max\left\{ \overline{f} \ \Big| \ \text{Ch}\left\{ \dfrac{v^{*T} \tilde{y}_0}{u^{*T} \tilde{x}_0} \geq \overline{f} \right\} \geq 1 - \alpha \right\} \geq \max\left\{ \overline{f} \ \Big| \ \text{Ch}\left\{ \dfrac{v^T \tilde{y}_0}{u^T \tilde{x}_0} \geq \overline{f} \right\} \geq 1 - \alpha \right\} \quad (6.9)$$

for any feasible solution (u, v).

Optimistic Ranking Criterion: The greater the optimal objective value is, the more efficient DMU_0 is ranked.

6.3.3 Maximal Chance Ranking Criterion

Sometimes the decision maker may want to maximize the chance of satisfying the event $\dfrac{v^T \tilde{y}_0}{u^T \tilde{x}_0} \geq 1$. In order to model this type of decision system, Liu [4–6] provided the dependent-chance programming (DCP). Here we carried out the DCP model into the hybrid DEA as follows:

$$\begin{cases} \theta = \max_{u,v} \text{Ch}\left\{ \dfrac{v^T \tilde{y}_0}{u^T \tilde{x}_0} \geq 1 \right\} \\ \text{subject to:} \\ \quad \text{Ch}\left\{ v^T \tilde{y}_k \leq u^T \tilde{x}_k \right\} \geq \alpha, \ k = 1, 2, \cdots, n \\ \quad u \geq 0 \\ \quad v \geq 0 \end{cases} \quad (6.10)$$

in which $\alpha \in (0.5, 1]$.

6.3 Hybrid DEA Ranking Criteria

Definition 6.5. A feasible solution (u^*, v^*) is called a maximal chance optimal solution to the hybrid programming model (6.10) if

$$\text{Ch}\left\{\frac{v^{*T}\tilde{y}_0}{u^{*T}\tilde{x}_0} \geq 1\right\} \geq \text{Ch}\left\{\frac{v^T\tilde{y}_0}{u^T\tilde{x}_0} \geq \overline{f}\right\} \tag{6.11}$$

for any feasible solution (u, v).

Maximal Chance Ranking Criterion: The greater the optimal objective value is, the more efficient DMU_0 is ranked.

6.3.4 Hurwicz Ranking Criterion

Similar to Sect. 6.2, the pessimistic model, which considers the total distances to an inefficient frontier, can be given as

$$\begin{cases} \theta_2 = \max \sum_{i=1}^{p} s_i^- + \sum_{j=1}^{q} s_j^+ \\ \text{subject to:} \\ \quad \text{Ch}\left\{\sum_{k=1}^{n} \tilde{x}_{ki}\lambda_k \geq \tilde{x}_{0i} + s_i^-\right\} \geq \alpha, \quad i = 1, 2, \cdots, p \\ \quad \text{Ch}\left\{\sum_{k=1}^{n} \tilde{y}_{kj}\lambda_k \leq \tilde{y}_{0j} - s_j^+\right\} \geq \alpha, \quad j = 1, 2 \cdots, q \\ \quad \sum_{j=1}^{n} \lambda_j = 1 \\ \quad \lambda_j \geq 0, \quad j = 1, 2, \cdots, n \\ \quad s_i^- \geq 0, \quad i = 1, 2, \cdots, p \\ \quad s_j^+ \geq 0, \quad j = 1, 2 \cdots, q. \end{cases} \tag{6.12}$$

Definition 6.6. DMU_0 is α-inefficient if s_i^{-*} and s_j^{+*} are zero for $i = 1, 2, \cdots, p$ and $j = 1, 2 \cdots, q$, where s_i^{-*} and s_j^{+*} are optimal solutions of (6.12).

Above mentioned two models are both extreme cases: one is too optimistic and the other is too pessimistic. Thus, we employ the Hurwicz criterion, suggested by Leonid Hurwicz [2, 3] in 1951, which incorporates a measure of both by assigning a certain percentage weight λ to θ_1^* and $1 - \lambda$ to $-\theta_2^*$, $0 \leq \lambda \leq 1$:

$$\theta^* = \lambda \theta_1^* - (1 - \lambda)\theta_2^*. \tag{6.13}$$

The greater the ranking value θ^* is, the less efficient DMU_0 is ranked.

6.4 Hybrid Congestion

This section will introduce some congestions with uncertain random inputs and outputs. After identifying that DMU_0 is inefficient proposed by model (6.1), the following hybrid model is utilized:

$$\begin{cases} \max \sum_{i=1}^{m} \delta_i^- \\ \text{subject to:} \\ \quad \text{Ch}\left\{\sum_{k=1}^{n} \tilde{x}_{ki}\lambda_k - \delta_i^- \geq \tilde{x}_{0i} - s_i^{-*}\right\} \geq \alpha, i = 1, 2, \cdots, p \\ \quad \text{Ch}\left\{\sum_{k=1}^{n} \tilde{y}_{kj}\lambda_k \geq \tilde{y}_{0j} + s_j^{+*}\right\} \geq \alpha, \quad j = 1, 2 \cdots, q \\ \quad \sum_{k=1}^{n} \lambda_k = 1 \\ \quad \lambda_k \geq 0, \quad k = 1, 2, \cdots, n \\ \quad s_i^{-*} \geq \delta_i^-, \quad i = 1, 2 \cdots, p \end{cases} \quad (6.14)$$

in which s_i^{-*} and s_j^{+*} are optimal solutions of model (6.1).

Finally, the congesting amount in the total slack associated with s_i^{-*} is defined

$$s_i^{-c*} = s_i^{-*} - \delta_i^{-*}, \ i = 1, 2, \cdots, p. \quad (6.15)$$

Therefore, we have the following theorem on α-inefficient and congestion:

Theorem 6.1. *At an optimum of (6.1), (6.14) and (6.15), we have the following:*

(a) *If there exists at least one $s_i^{-*} > 0$, $s_j^{+*} > 0$, $1 \leq i \leq p$, $1 \leq j \leq q$, then DMU_0 is α-inefficient.*
(b) *If there exists at least one $s_i^{-c*} > 0$, $1 \leq i \leq p$, then DMU_0 is α-inefficient, and congestion is present.*
(c) *If $s^{+*} = 0$ and $s^{-c*} = 0$, then DMU_0 is on a segment of the hybrid frontier.*

References

1. Charnes A, Cooper W (1961) Management models and industrial applications of linear programming. Wiley, New York
2. Hurwicz L (1951) Optimality criteria for decision making under ignorance. Cowles Commission discussion paper, Chicago, 370

3. Hurwicz L (1951) Some specification problems and application to econometric models (abstract). Econometrica 19:343–344 (1951)
4. Liu B (1997) Dependent-chance programming: a class of stochastic programming. Comput Math Appl 34(12):89–104
5. Liu B (1999) Dependent-chance programming with fuzzy decisions. IEEE Trans Fuzzy Syst 7:354–360
6. Liu B (2002) Random fuzzy dependent-chance programming and its hybrid intelligent algorithm. Inf Sci 141(3–4):259–271

List of Frequently Used Symbols

Cr	Credibility measure
$(\Theta, \mathcal{P}, \mathrm{Cr})$	Credibility space
μ, ν	Membership function
(a, b)	Equipossible fuzzy variable
(a, b, c)	Triangular fuzzy variable
(a, b, c, d)	Trapezoidal fuzzy variable
$\xi_{\sup}(\alpha)$	α-optimistic value of Fuzzy variable ξ
$\xi_{\inf}(\alpha)$	α-pessimistic value of Fuzzy variable ξ
$\mathcal{M}$	Uncertain measure
$(\Gamma, \mathcal{L}, \mathcal{M})$	Uncertainty space
Φ, Ψ, Υ	Uncertainty distributions
$\Phi^{-1}, \Psi^{-1}, \Upsilon^{-1}$	Inverse uncertainty distributions
$\mathcal{L}(a, b)$	Linear uncertain variable
$\mathcal{Z}(a, b, c)$	Zigzag uncertain variable
$\mathcal{N}(e, \sigma)$	Normal uncertain variable
$\mathcal{LOGN}(e, \sigma)$	Lognormal uncertain variable
E	Expected value
V	Variance
Pr	Probability measure
Ch	Chance measure
$\emptyset$	The empty set
$\Re$	The set of real numbers
$\vee$	Maximum operator
$\wedge$	Minimum operator
$\forall$	Universal quantifier
$\tilde{x}$	Input vector
$\tilde{y}$	Output vector
u	Vector of input weights
v	Vector of output weights
$\tilde{X}$	Input matrix
$\tilde{Y}$	Output matrix

MIX
Papier aus verantwortungsvollen Quellen
Paper from responsible sources
FSC® C105338

If you have any concerns about our products,
you can contact us on
ProductSafety@springernature.com

In case Publisher is established outside the EU,
the EU authorized representative is:
Springer Nature Customer Service Center GmbH
Europaplatz 3, 69115 Heidelberg, Germany

Printed by Libri Plureos GmbH
in Hamburg, Germany